THE QUEST FOR THE CURE

THE QUEST FOR THE CURE

THE SCIENCE AND STORIES BEHIND THE NEXT GENERATION OF MEDICINES

BRENT R. STOCKWELL

COLUMBIA UNIVERSITY PRESS
NEW YORK

Columbia University Press
Publishers Since 1893
New York Chichester, West Sussex

Library of Congress Cataloging-in-Publication Data
Stockwell, Brent R., author.
The quest for the cure : the science, stories, and struggles behind
the next generation of medicines / Brent R. Stockwell, PhD.
p. ; cm.
Includes bibliographical references.
ISBN 978-0-231-15212-9 (cloth : alk. paper)—
ISBN 978-0-231-52552-7 (ebook)
1. Drug development. 2. Drugs—Design. 3. Proteins—
Metabolism. I. Title.
[DNLM: 1. Drug Discovery—methods. 2. Drug Design.
3. Proteins—chemistry. 4. Proteins—metabolism. QV 744]
RM301.25.S76 2011
615′.19—dc22 2010053573

Columbia University Press books are printed on permanent and durable
acid-free paper.
This book is printed on paper with recycled content.
Printed in the United States of America

c 10 9 8 7 6 5 4 3 2

TO JONAH AND ARI,

THE NEXT GENERATION

HOW MANY THINGS, TOO, ARE LOOKED UPON AS QUITE IMPOSSIBLE UNTIL THEY HAVE BEEN ACTUALLY EFFECTED?

—PLINY THE ELDER

CONTENTS

LIST OF ILLUSTRATIONS

Note: Plates follow page 140

PREFACE

There is a shortage of new medicines. This is a manifestation of the fact that it is becoming increasingly difficult to discover new drugs. One disturbing possibility is that we are approaching the end of pharmaceutical medicine. Although this sounds surprising, there is substantial evidence to support this hypothesis. New drug approvals have been declining for several years. The recent large-scale reordering of the pharmaceutical industry to fill shrinking drug pipelines, involving acquisitions of Wyeth by Pfizer, Schering-Plough by Merck, and Genentech by Roche, is a product of this decline.

Some researchers believe it will still be possible to create a wealth of powerful new drugs, but a particular scientific challenge has to be overcome first. According to this view, the answer to curing intractable diseases lies in finding the right kinds of drug molecules, the ones that can interact with proteins that are said to be "undruggable," or intractable. *The Quest for the Cure* is about this high-stakes search by scientists to create the next generation of medicines.

Part One, involving the first six chapters, describes the current problem associated with creating new medicines—that so few proteins are considered druggable. In other words, there is no hope today of making drugs that interact with most proteins. To understand this central challenge in discovering new drugs, Part One looks back at the history of how existing drugs were created and tells the stories of the science and scientists who brought us to where we are today. Part One concludes with the idea that new scientific approaches are needed to solve the druggability problem.

Part Two describes emerging approaches for tackling the problem of the undruggable proteins and some of the leading scientists involved in these efforts. This second half of the book raises the hope that the major challenge facing drug hunters is not ultimately insurmountable and that it might be overcome by the approaches described.

To understand the narrative in this story, it is necessary to understand some particular scientific details. I have endeavored to make this science broadly accessible, but without compromising accuracy. There is a glossary at the end of the book to explain technical terms. Although these terms are explained in the main text when they are introduced, readers may find it helpful to turn to the glossary from time to time.

It is worth noting that gene and protein abbreviations are used here in capital letters, as opposed to the conventional scientific style, which uses different formats for mouse and human genes and proteins. It is hopefully simpler for the reader to use the same style, irrespective of whether we are speaking about a human or mouse gene or protein. I would also like to apologize to the many researchers whose work I did not have space to cite. There is a wealth of relevant research to discuss, but I selected particular themes that I thought created a coherent narrative.

In summary, in this book I have sought to address the question: What kinds of molecules will become the next generation of medicines, and which scientists are on the path towards making these breakthroughs? *The Quest for the Cure* describes the possibilities for creating drugs that interact with the undruggable proteins, thereby addressing currently incurable diseases. Along the way the book describes the lives and discoveries of the many researchers who have been involved in this major challenge of our time. This is a story about a grand effort by many researchers, and I invite you to share their triumphs and struggles as you join them on their quest.

ACKNOWLEDGMENTS

Writing a book for the general public has been a goal of mine for many years. This book represents the culmination of a long thought process that involved many people, and I am grateful for the help I have received on this journey. First I want to thank my two children, Jonah and Ari, for helping me think up a suitable title, and my wife, Melissa, for poring over the chapters as I wrote them, making valuable suggestions. I also want to thank Michael Scharf, who, during a long car ride to New York City, convinced me to finally undertake this project, and then proceeded to give me advice throughout the evolution of the book.

I am grateful to my mother and stepfather, Serena Stockwell and Dennis Schuman, who have encouraged me in this project and offered much feedback—it certainly helps to have an editor as a mother. My wife's parents, Constance and Robert Scharf, have assisted me in explaining complex science in simple terms. I am especially grateful to Constance for scrutinizing my first complete draft and providing many specific comments.

I appreciate the efforts of my editor, Patrick Fitzgerald, who helped me navigate the new (to me) world of book publishing and offered key insights in translating complex science into everyday language, and Stuart Firestein, who introduced me to Patrick and supported my effort from the start. I was also fortunate to get needed help from my assistant, Jessica Kunkler.

I am indebted to a number of scientists who took time out of busy schedules, on short notice, to read parts of the book along the way and to offer expert critiques. These generous individuals include Brian Druker, Solomon Snyder, Eric Lander, Stephen Fesik, George Prendergast, Dan Von Hoff, James Crow, John Hunt, Arthur Palmer, James Wells, Anthony Czarnik, Jonathan Ellman, Robert Weinberg, Peter Schultz, Michael Foley, Paul Wender, Sam Gellman, Ronald Bres-low, and Stuart Schreiber.

Finally, I am grateful to you, the reader, for taking a chance on this book. I hope I was able to convey the promise and excitement at this time in biomedical research, and to share some of the big questions that motivate scientists in this area.

ABBREVIATIONS

ACE: angiotensin-converting enzyme

AIDS: acquired immunodeficiency syndrome

ALS: amyotrophic lateral sclerosis, a neurodegenerative disease

ATP: adenosine triphosphate, a molecule used for storing and providing energy in cells

BCR-ABL: The gene and protein found in patients with the Philadelphia chromosome

CEO: chief executive officer

CML: chronic myelogenous leukemia, a type of cancer

DNA: deoxyribonucleic acid, the chemical that stores the genetic information

ETH: the Swiss Federal Institute of Technology

FDA: Food and Drug Administration, the government agency responsible for approving drugs for use in the United States

GAP: guanosine triphosphatase activating protein, a type of protein that stimulates G proteins, such as RAS, to convert GTP to GDP

GDP: guanosine diphosphate, a molecule that can bind to G proteins such as RAS and turn off their signaling functions

GPCR: G-protein-coupled receptor, a type of protein on the surface of cells that is targeted by many existing drugs

GTP: guanosine triphosphate, a molecule that is a building block for RNA and also turns on G proteins such as RAS
HCNO: the molecular formula of fulminic acid
HIV: human immunodeficiency virus
HNCO: the molecular formula of isocyanic acid
IL-2: interleukin 2, a protein that controls the immune response
IL-2R: the interleukin 2 receptor, found on the surface of cells
LHRH: luteinizing-hormone-releasing hormone
LogP: logarithm of partition coefficient, a measure of hydrophobicity, or greasiness
MCL: mantle cell leukemia
MDM2: murine double minute-2, a cancer-causing protein that inactivates the tumor suppressor protein p53
NMR: nuclear magnetic resonance
NSAID: non-steroidal anti-inflammatory drug, such as aspirin
PAP: phenylaminopyrimidine
PDE: phosphodiesterase, an enzyme involved in breaking down the signaling molecule cyclic adenosine monophosphate (cAMP)
PTI: protein trypsin inhibitor
RAL: RAS-like protein
RALGDS: RAL guanine nucleotide dissociation stimulator
RAS: a gene mutated in cancers, originally found in a rat sarcoma virus (includes K-RAS, N-RAS, H-RAS)
RHEB: a protein related to RAS, short for "RAS homolog enriched in brain"
RNA: ribonucleic acid
RSV: Rous sarcoma virus
SRC: a cancer-causing gene found in the Rous sarcoma virus
SV40: simian virus 40
TGF-beta: transforming growth factor beta
VC: venture capitalist
VEGF: vascular endothelial cell growth factor

THE QUEST FOR THE CURE

PART ONE

THE VANISHING CURES

1

THE DRUG DISCOVERY CRISIS

One recent morning I received an e-mail message with the subject line "Sad News." As I clicked on the message title and the full message popped up, I expected to hear of the passing of a distant colleague, someone whose retirement party I might have attended long ago as a graduate student, or a senior emeritus faculty member I might have met when I was in college. Instead, I was shocked to learn that my friend Darin had died of cancer at age 36.

I soon learned that Darin had been diagnosed with liver cancer some months before. He had undergone chemotherapy, without success. He had a transplant operation and had been recuperating at home, feeling somewhat better. Ultimately, the transplant didn't take, and he began to get spiking fevers and became extremely ill. A month earlier he had gone into a hospice, and he then passed away one quiet Thursday, the day I received the e-mail message.

When we were 18, Darin and I spent a year traveling around Israel together on an organized program; we used to debate such minutiae as the proper number of times to reuse a razor blade. I hadn't spoken to him in years, but the news of his sudden death hit me unexpectedly hard.

Darin was exuberant and friendly, with a touch of a caustic wit. He invariably left me smiling to myself after a round of informal debate.

Prior to his diagnosis Darin had been just reaching his stride, personally and professionally. As someone who has done research into the biology of cancer, I could imagine the lethal progression of his disease and the fear and anger he must have felt as the available treatments failed him. His life was ultimately left unfinished because of a small number of aberrant cells in his liver and a lack of effective drugs. With these somber thoughts, I groped for the meaning of Darin's untimely passing.

Darin's tragically short life is not unique. My cousin Daniel was 35 when he was dining downtown in New York City on a summer evening with his girlfriend; he suddenly collapsed from a seizure. He was rushed to the hospital, where he learned he had an extremely aggressive form of brain cancer. He succumbed within a year, causing a painful void in the lives of his friends and family.

Cancer is only one of many diseases that can appear unexpectedly. Lou Gehrig, the tenacious baseball player, was diagnosed at age 36 with a deadly form of nerve degeneration that would paralyze him and then claim his life. Woody Guthrie, the singer of mesmerizing folk ballads, succumbed to Huntington's disease, which causes constant, involuntary dance-like movements and is ultimately fatal. Ronald Reagan, a former U.S. president, developed Alzheimer's disease, leading to the erasure of his life's memories.

THE SHRINKING NUMBER OF NEW MEDICINES

Vast hordes of researchers are working to find cures for nearly every disease known to medicine. With these keen intellects working diligently, competing fiercely to solve these mysteries, you would expect that no stone has gone unturned, and no avenue unexplored. Yet we have not found cures for most types of cancer. Other diseases, including Alzheimer's, Huntington's, Lou Gehrig's, and Parkinson's, remain depressingly void of curative therapies. Why?

The United States and the rest of the world are facing a drug discovery crisis, a fact that has been evident to researchers for the

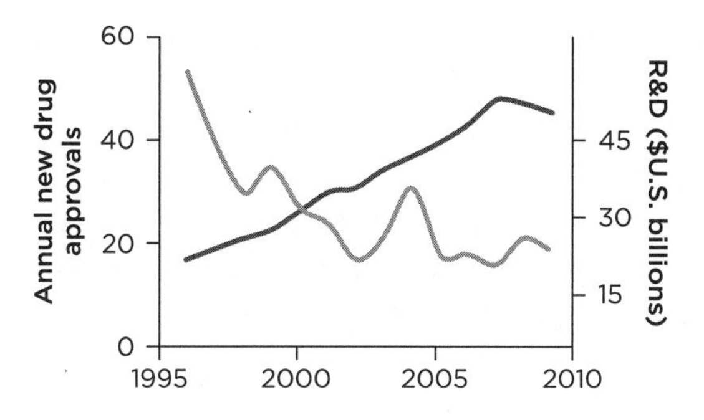

Figure 1.1 The shrinking number of new medicines. Shown are the number of new drugs approved (defined as new chemical entities, NCEs) by the U.S. Food and Drug Administration (gray line) and the amount of research and development funding by the U.S. pharmaceutical industry each year from 1996 to 2009 (black line). Despite a significant increase in research funding, the number of new drugs has plummeted, indicating a fundamental barrier exists to creating new medicines. *Data sources*: Pharmaceutical Research and Manufacturers of America, *Pharmaceutical Industry Profile 2010* (Washington, DC: PhRMA, March 2010), 26; Hughes, B., 2009 FDA drug approvals. *Nat Rev Drug Discov* 9, 90 (2010).

last decade. The number of new drug approvals each year has declined more than 50% over this time, despite a massive increase in the amount of research funding devoted to drug discovery research. The cost and time needed for making each new drug is increasing dramatically, costing in excess of a billion dollars and taking more than a decade of late-stage research. Most importantly, researchers are finding fewer new drugs (see Figure 1.1).

Despite the common perception of a boundless frontier, medical progress, measured in the form of new drug approvals, is in fact slowing.[1] The pharmaceutical industry is in disarray, suffering from meager drug pipelines and laying off many employees. If the trend continues, patients face a bleak future in which dwindling progress will be seen against disease. We may have to adjust our expectations to that of a future in which many people inevitably succumb to tragic and painful diseases at all ages, with no substantial progress being made in creating new medicines.

This problem can be traced, at least in part, to an insurmountable obstacle that has stymied those who search for new medicines—the challenge of the *undruggable proteins*. These are the proteins that cannot be affected by drugs; these proteins, it is argued, lie beyond the reach of human ingenuity. These proteins are the ones that no drug has ever tamed.

WHAT IS A DRUG?

To understand the nature of these undruggable proteins and how they might ultimately be used to make medicines, it is crucial to understand what drugs are and how they function. When you swallow a pill, the drug inside that pill gets absorbed into your bloodstream, where it is distributed to the tissues in your body (see Figure 1.2). Many times, the drug is able to slip inside cells that make up these tissues and spread throughout the interior of cells, much like when you add a drop of ink into a cup of water.

The interior of a cell is made up of thousands of different proteins, which carry out most of the work of the cell. Proteins perform specific functions, like producing energy for the cell or enabling the cell to divide into two new cells. Such cellular processes require the coordinated activities of many different proteins. In terms of physical shape these proteins look like beautifully molded pieces of clay, each uniquely sculpted to suit its function. Some are long and thin, others are short and squat. Some have large holes in their middle, resembling a doughnut. Moreover, the surfaces of some proteins are smooth and featureless, while others are craggy and filled with crevices. In biology structure implies function, and this is true of proteins. Each protein has a shape that allows it to carry out its unique function within the cell.

Drugs act by changing the function of a protein. To interfere with the function of a protein, a drug has to attach itself to a protein, fitting snugly into a crevice on the surface (see Figure 1.3).[2]

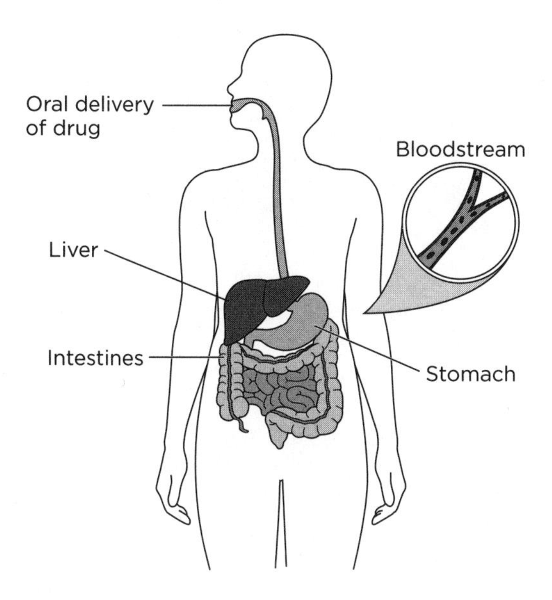

Figure 1.2 How a drug is distributed through an organism. When a drug is taken by mouth, it enters the gastrointestinal tract (including the stomach), is absorbed through the intestines, and enters the bloodstream. From there it is distributed to the different tissues and organs of the body in a way that depends on the specific molecular structure of the drug. In addition, as the blood passes through the liver, the drug gets chemically modified in a process called metabolism, which usually results in the drug being excreted through the kidneys into urine.

How does a drug stick to a protein and alter its function? Protein molecules are large, whereas drugs are much smaller than proteins. If a protein were the head of a life-size statue, a drug would be like an earplug sitting in the ear canal of the statue. Because of this difference in size, researchers call these drugs "small molecules"—they are small compared with proteins. Therefore, discovering a drug requires finding a small molecule that can stick to a specific protein and change its function. Such a protein is said to be the target of the drug.

Drugs stick to proteins when they have complementary shapes and properties. Think of putting a key in a lock or a square peg in a square hole. The shape of a drug must match the shape of a small

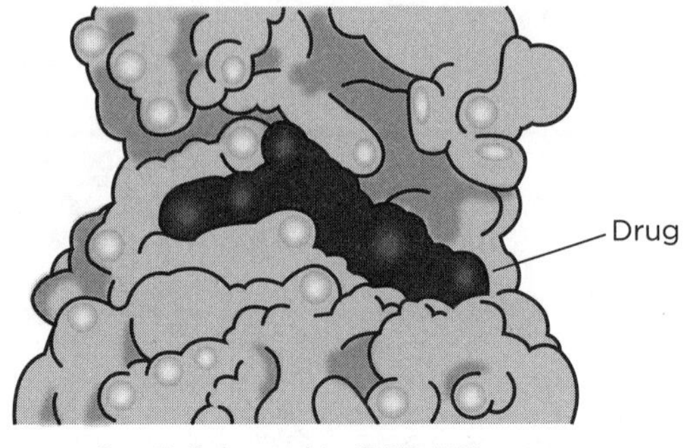

Figure 1.3 Example of a drug binding to a protein. The drug imatinib (black) is shown interacting with the BCR-ABL oncoprotein. The BCR-ABL protein is shown in gray in a space-filling model.

indentation, or pocket, on the surface of a protein in order for the drug to stick tightly to the protein. This tight-fitting interaction between a drug and a protein is called *binding*. Yet shape matching is not enough for binding. The drug must also have the right amount of hydrophobicity (hai-droh-foh-BISS-ih-tee), or greasiness, in the right places. The drug must also have electrical charge and hydrogen bonds, or stickiness, in just the right positions. Moreover, a drug may cause the shape of a protein to change upon binding, like pushing a spoon into a soft clay sculpture. There are other, more complicated aspects of drug binding. For now, suffice it to say that it is generally difficult to find drugs that will bind tightly to a specific protein.

THE UNDRUGGABLE PROTEINS

Here is the surprising fact: All of the 20,000 or so drug products that ever have been approved by the U.S. Food and Drug Administration interact with just 2% of the proteins found in human cells.[3] This means that the vast majority of proteins in our cells—many of which, in theory, can modulate disease processes—have never been targeted before with a drug.

Two responses to this news are possible. On one hand, you might be elated to discover that there is a huge reservoir of proteins that hasn't been mined for new drugs and therefore be optimistic about the future of drug discovery and medicine. On the other hand, you might be pessimistic because you suspect that after decades of struggle, we have not found drugs that are able to bind to these proteins. In other words, perhaps the 2% of proteins that have been targeted with drugs are the only ones capable of being targeted.

What do the data tell us? Unfortunately, there is a fair amount of evidence suggesting that most proteins do not readily interact with drugs and that the more pessimistic scenario might be the correct one. For example, in a recent study at the pharmaceutical company GlaxoSmithKline, researchers tested up to 530,000 different small molecules against 67 typical proteins in a series of screens. A *screen* is a large-scale experiment that involves testing many different small molecules—in this case, 530,000 small molecules were tested one by one to see if any of them can stick to any of these proteins.[4]

None of these 67 screens of 530,000 compounds resulted in a small molecule drug that could be developed for use in patients. Many such experiences have led some researchers to the conclusion that most proteins are undruggable—they are simply not capable of binding in a selective way to small molecules of the type that can become drugs.

Perhaps the molecules that were tested in these screens simply weren't the right ones for the job. After all, even several hundred thousand molecules is an infinitesimal fraction of the number of possible small molecules that could be made in theory. This number of possible small molecules has been estimated to be on the order of 10^{60}; that is "1" followed by 60 zeros. Thus, even testing a million compounds is not that many in terms of sampling a significant fraction of all possible small molecules. To get a sense of how big the number 10^{60} is, if all the possible drug molecules filled up the space of 10,000 different planets, then 1 million chemicals would be represented by just a single molecule on a single one of these planets. All of the remaining molecules on all of the planets would still remain to be tested.

Nonetheless, researchers are only human. If you have a negative experience at something, you are less likely to try it again. If you have 67 negative experiences at screening small molecules, you are going to try hard to make the 68th experience a positive one. Thus many researchers have become risk averse, mainly in regards to selecting proteins to design drugs against. They do not want to pick an undruggable protein to study and end up failing to find a small molecule inhibitor.

I know how powerful this aversive conditioning can be. When I was a graduate student, I devised a screen to search for small molecules that could block the action of a specific protein, called TGF-beta (for transforming growth factor beta), which is involved in the formation of certain tumors. I developed a rapid test that would allow me to run this screen efficiently.[5] I then tested 16,000 synthetic small molecules and was deflated when I found none that were able to block the effects of TGF-beta on cells.

Not to be deterred, I decided to try testing some additional small molecules. This time, I tested different types of molecules—a set of 200 extracts obtained from marine sponges, provided to me by Phillip Crews. Phil and his colleagues travel the world to exotic locales, dive deep into the ocean to find unusual marine sponges, and then purify small molecules from these sponges using an extraction procedure.[6] In this procedure the sponge is mixed with an organic solvent—similar to the one used in dry cleaning clothes—to separate small molecules within the sponge from other material.

Each of these extracts typically has several dozen different small molecules together in a mixture. I tested these 200 extracts, not expecting to find anything. After all, if I didn't find any active molecules from among 16,000 synthetic chemicals, what was the chance I would find an active one in only 200 natural product extracts? I was surprised when one of these 200 extracts had a dramatic and striking ability to completely block the effects of TGF-beta in my test. I repeated this result several times to be sure. This is one of the first things researchers like to do when they get an exciting result—try to reproduce it a number of times to make sure it is real. Judah Folkman (who was both a surgeon and a basic scientist) once said that the difference between surgeons and basic scientists is that when someone can't reproduce the results of a basic scientist, the scientist becomes alarmed. When someone can't reproduce the results of a surgeon, the surgeon takes it as a compliment to their superior skill.

After confirming the result, I worked with Phil and his lab members to purify the single compound that was responsible for this striking activity. We went through several rounds of purification until we arrived at an active mixture of two or perhaps three chemicals, and we knew that one last experiment should separate them from each other. This was the pivotal moment, and I remember my anticipation. I was about to discover a powerful new natural product.

Despite my excitement, I carried out the last test carefully and methodically, wanting to be 100% sure of the result. The way I determined the effect of TGF-beta on these cells involved measuring the amount of cell proliferation occurring by placing a piece of film on a plastic dish containing the cells and then looking to see whether a dark spot appeared on the film. When I treated the cells with TGF-beta, the dark spot would disappear. The natural product I was chasing caused the dark spot to reappear, thus blocking the effects of TGF-beta on these cells.

After doing this last experiment, I went into the film development room and laid the film down onto the dish containing the cells. I carefully fed the film into the automated developer, and waited. Seconds ticked by as I anxiously awaited the results. Finally, a piece of film gently slid from the developer. I held the film up to the light and stared at it. It was completely blank.

Somehow, none of the samples from the final purification were active. I repeated the experiment again to be sure, but the result was the same. I was disappointed, but I assumed we would repeat the whole process with a new batch of this sponge from deep in the ocean. However, when I told my collaborators the result, they sighed and said, well, that is the end of that. That was the world's supply of the sponge, and they had no way of getting any more.

This experience led me to avoid working with natural product extracts in the future. Although it is always possible that the next experiment might have a different outcome, the time and resources expended on this effort convinced me to be more risk averse in terms of working with extracts of marine sponges, because although they can do remarkable things, they are not a renewable resource—you can run out of the material in the middle of an experiment. Thus I can empathize with the perspective of those who wish to work only on the most tractable drug targets.

There is a second message in this story. The 16,000 synthetic small molecules I tested initially were fairly simple molecules that were assembled quickly and easily in commercial labs. These chemicals lacked architectural complexity. They were small, flat, and didn't have a great deal of functionality. Surprisingly, many of the molecules tested by drug discovery labs are such simple compounds assembled in trivial ways by commercial vendors.

In contrast, natural products are often more complex, larger, and have a greater variety of chemical functionality. While I don't know the specific chemical structure of that active molecule we failed to purify because we didn't have enough material, I find it intriguing that the result of screening natural product extracts was so different from that of screening synthetic molecules. If I had only screened synthetic molecules, I might have been tempted to conclude that my assay was undruggable and that it was physically impossible to find any molecules that could be active in this test.

Gun-shy researchers who have been burned working on difficult drug targets may be making a mistake in forgoing these proteins in their drug-hunting endeavors. It is surely easier to find drugs targeting the 2% of proteins that are known to be druggable, but will these drugs be maximally effective, and will they cure our most challenging ailments?

This is where some advice once provided to me by my stepfather, Dennis, is germane. At the time, I was preparing to start graduate school in Harvard's Chemistry Department, and I was concerned about finding an apartment. I had seen what I thought was a perfect studio in one of the graduate student housing complexes, but the rent was more than I could afford. I had mentally crossed this apartment off my list and was only considering rundown studios farther from campus, like the one in Somerville where the bed folded up into the wall to save space during the day.

At this point, Dennis remarked, "You are confusing two different questions. First, there is the question of what you want. Only then should you consider the second question, which is how do you find a path to get there." He went on to say that if I loved one apartment, then I should focus on finding a way to realize that goal rather than settling for something less desirable that would leave me unhappy. First, the goal. Then, the means. This straightforward view of the world is applicable to the problems facing academia and the pharmaceutical industry in discovering new drugs.

If the goal is to create the most effective cures for cancer, Alzheimer's, and other devastating diseases, then perhaps we should put all proteins on the table, so to speak, as possible drug targets. After settling on this goal, the global science community can try to figure out how to reach it.

To begin with, we must define this crucial term "druggable." Researchers use this term in the following sense: "Protein X is druggable, but protein Y is not druggable."[7] What they mean is that with a reasonable amount of time, effort, and money, it should be possible to find a small molecule that can bind to protein X. This small molecule is then likely to alter the ability of protein X to carry out its normal function in cells and might as a consequence cause the cells to behave differently, which in turn might change the course of a disease.

On the other hand, if protein Y is considered undruggable, this implies that it is difficult, if not impossible, to find small molecules that can bind to protein Y. How can we know this? Proteins fall into different families, depending on their shape. Some families of proteins are known to be suitable for binding to drugs, while for other families of proteins, no one has been able to find a drug that can bind to them. Even worse, for many of these putatively undruggable protein families, when researchers have tested hundreds of thousands of small molecule drug candidates for their ability to bind to them, none have been found, as in the Glaxo example. These proteins appear to be unstoppable, intractable, and thus, in the parlance of the pharmaceutical industry, undruggable. If these proteins are indeed undruggable, then the diseases they control are incurable.

However, if someone were to find a way to make drugs against some of these undruggable proteins, perhaps all of these undruggable proteins should be put back on the table.

THE EVOLUTION OF KINASES FROM UNDRUGGABLE TO DRUGGABLE

Kinases (KAI-nay-sez) are a powerful example of putatively undruggable proteins that in fact have been rendered druggable. Kinases are proteins that attach a small chemical group onto other molecules. Kinases are important in relaying messages from one location in a cell to another, among other functions. There are more than 600 different kinases in humans.

For many years, kinases were considered undruggable.[8] The argument for their undruggability was based on the nature of the chemical reaction that these kinase enzymes facilitate. The function of all enzymes is to catalyze, or accelerate, one or more chemical reactions in the cell. What does that mean? It means that an enzyme can make a chemical reaction occur quickly, even if that reaction normally occurs slowly or not at all. The enzyme brings reacting molecules together in a way that facilitates their reaction, like a molecular matchmaker. The specific chemical reaction catalyzed by kinases is the addition of a small phosphoryl (FAHS-for-eel) group onto another molecule, such as another protein.

A phosphoryl group is a cluster of four atoms—one phosphorus atom surrounded by three oxygen atoms. This group of four atoms, once attached to a protein, changes the properties of the protein. The phosphoryl group is negatively charged and also takes up some amount of space. This charge and bulk can have a dramatic effect on the properties of proteins and other molecules once it is attached to them, causing their normal functions to turn on or off or changing their interactions with other molecules.

The problem with trying to create drugs targeting kinases, it was believed, was that all kinases use the molecule adenosine triphosphate (ATP) as a source of the phosphoryl group. ATP is a naturally occurring small molecule in the cell that has a phosphoryl group attached to it, ready to be transferred to another molecule.

Since all kinases use ATP in a similar way, the thinking was that it would be impossible to design a drug to stick to the pocket on one kinase where ATP binds without sticking to the same pocket on all kinases, because that pocket is so similar in each member of this protein family. Inhibiting the functions of all 600 kinases in a cell at the same time would almost certainly cause severe and undesired drug side effects, so this was deemed a poor strategy for creating a drug. This was a persuasive argument for avoiding any attempt to create drugs that inhibit kinases.

Nonetheless, some researchers pushed forward with the dream of finding small molecules that could selectively bind one particular kinase. A group led by Nicholas Lydon at the pharmaceutical company Ciba-Geigy (pronounced SEE-bah GAI-gee and now called Novartis) reported that a specific kind of small molecule, a phenylaminopyrimidine (FEN-yl a-MEEN-oh pu-RIM-a-deen, abbreviated PAP) could bind to the kinase named protein kinase C.[9]

This initial observation was still not useful, as PAP was not selective—it bound to many different kinases. Nonetheless, Lydon and his colleagues persevered and tried making a series of similar-looking molecules with slight variations to them. The thinking was that by keeping the core of the molecule the same, these new PAP derivatives would still be able to bind to kinases. However, by slightly changing the chemical groups decorating the core of the molecule, it might be possible to find a PAP-like molecule that could bind to one kinase but not to others.

This approach ultimately bore fruit, described below, in the discovery of imatinib (ih-MAH-teh-nib), also known as Gleevec (GLEE-vek). Imatinib is a small molecule, built up from 29 carbon atoms, 31 hydrogen atoms, 7 nitrogen atoms, and 1 oxygen atom. Atoms form the basic building blocks of matter. When atoms are connected to each other in particular ways, specific molecules are formed with unique, and often unexpected, properties. The connections between these atoms are precise—if the atoms are connected differently, a different molecule is formed with different medicinal properties. Small molecules like imatinib look like Tinkertoys, in which small spheres are connected with wooden sticks. The sticks represent the bonds, and the spheres represent the atoms.

Imatinib binds to a small number of kinases inside cells. One of the kinases targeted by imatinib is BCR-ABL (described below), which is present in patients with a specific type of cancer called chronic myelogenous leukemia (CML).[10] This relatively rare disease turned out to be a battleground that forged the modern approach to cancer drug discovery.

CHRONIC MYELOGENOUS LEUKEMIA AND THE BCR-ABL KINASE

Patients with CML have an abnormally large number of white blood cells of a specific type called *granulocytes*, so named because they look granular under a microscope.[11] These abnormal granulocytes continue to replicate themselves over and over without limit, leading to their accumulation in patients with CML.

In most patients the defect can be traced back to a specific problem in the genetic instructions in these cells, found in their chromosomal DNA. These patients have a characteristic chromosomal rearrangement known as the Philadelphia chromosome, named for the city in which it was discovered in 1960. This seminal finding was made by Peter Nowell and David Hungerford. Peter Nowell trained as a pathologist after medical school, when he began studying human leukemias. He worked for two years in the Navy on blood cells, and then started his own research program at the University of Pennsylvania as a new faculty member in 1956. He began working with David Hungerford, who was a graduate student at Fox Chase Cancer Center, as both had been developing an interest in studying chromosomes in leukemias. In examining the chromosomes present in leukemia cells, they observed a tiny, unnatural chromosome that was present specifically in leukemia cells derived from patients with CML. This tiny mutant chromosome became known as the Philadelphia chromosome.[12]

All the genetic instructions for the functioning of a cell are found in the nucleus in the form of a series of DNA molecules called *chromosomes*. There are 46 chromosomes present in normal cells in humans. The Philadelphia chromosome is an abnormal chromosome in which a fragment of chromosome 9 exchanges places with a fragment on chromosome 22. Perhaps not surprisingly, this causes a problem: there is a gene found right at the site of chromosomal breakage. When these two chromosome fragments exchange with each other, a new gene is created at the junction point. The new hybrid gene, which is a combination of the BCR gene found on chromosome 22 and the ABL gene on chromosome 9, is called, appropriately enough, BCR-ABL (see Figure 1.4).[13] This fusion gene encodes faulty instructions that cause excessive cell division.

The ABL gene tells cells how to make the ABL protein, which in turn causes some types of cells to divide. Once the ABL protein is turned off, the cells stop dividing. However, the BCR-ABL fusion protein present in patients with CML is not regulated in this usual way. Instead, the BCR-ABL protein is always "on," telling cells to continue replicating themselves by dividing rapidly without end. This is why CML patients develop a leukemia in which granulocytes are overabundant—the granulocytes and their precursors are driven to divide endlessly because of this unnatural fusion gene

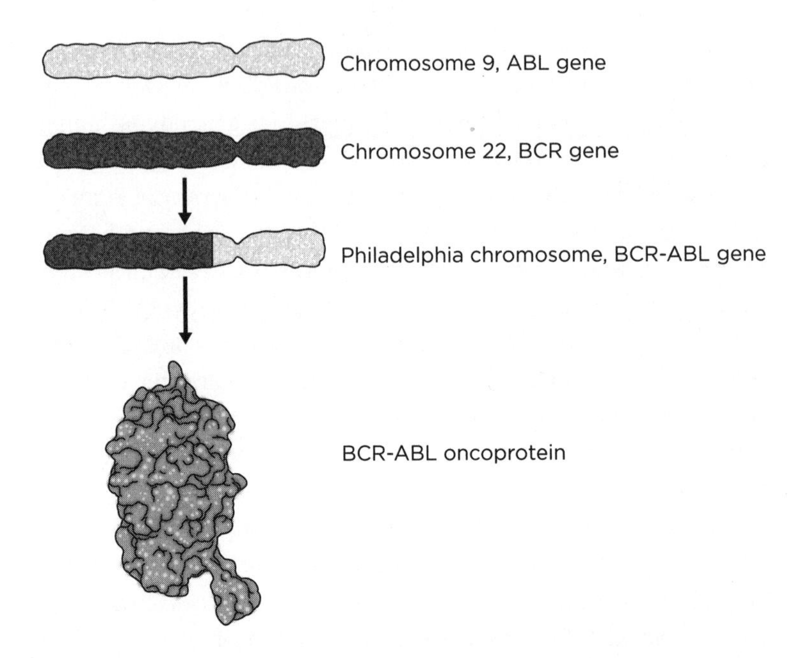

Figure 1.4 The Philadelphia chromosome and BCR-ABL. Patients with chronic myelogenous leukemia (CML) have tumor cells containing an aberrant chromosome fusion event, known as the Philadelphia chromosome. This chromosome joins parts of the normal chromosomes 9 and 22 to create a new gene at the site where the two chromosomes join together that encodes for the cancer-causing BCR-ABL oncoprotein.

on the Philadelphia chromosome. There is a silver lining, from the perspective of drug therapy. When the mutant BCR-ABL fusion protein is produced in these patients, the leukemia cells that result become dependent on mutant BCR-ABL for their survival. This is somewhat akin to drug addiction in humans—someone who has never taken an illicit drug such as heroin might be able to try it once, not take it again, and not suffer from any withdrawal symptoms, while people who take heroin regularly become dependent on the continued presence of the drug in their bodies. If such addicts suddenly stop taking heroin, they begin to experience painful physical changes including insomnia, diarrhea, vomiting, cold flashes, and muscle pain. We refer to this continued dependence on the presence of a drug as an *addiction*. The user of a drug such as heroin becomes *addicted* to the drug.

Bernard Weinstein pioneered the concept of "oncogene addiction"—that tumor cells become addicted to mutated proteins.[14] This idea is now widely cited and accepted. Recently Bernie sent me a handwritten note, along with two of his papers on the subject, after I had published a related article. I thought his handwritten note was elegant, in the age of e-mail and text messaging. I replied that I would enjoy getting together to talk more about our research; we agreed to meet. Tragically, shortly after this, Bernie passed away. He had been a pioneer in the field of cancer research.

The concept of oncogene addiction is crucial for understanding how a drug like imatinib is so effective against tumor cells. In the same way that a normal person can try heroin once and then stop without experiencing serious physical problems, blocking the activity of some proteins in normal cells doesn't cause a serious problem. However, just like the case of a heroin addict who stops taking heroin, blocking the function of the same proteins in cancer cells can cause serious problems and eventually cell death if the cancer cell has become addicted to the mutant protein.

Thus a drug that blocks BCR-ABL can selectively kill leukemia cells in patients with CML, because those leukemia cells have become addicted to BCR-ABL. A drug like imatinib can enter these leukemia cells, bind to the BCR-ABL protein, and prevent it from carrying out its activities in the cell. This leads to cellular symptoms of withdrawal, eventually resulting in the death of these leukemia cells.

In 2009 the Lasker-DeBakey Clinical Medical Research Award (colloquially known as the Lasker Award, and often a prelude to the Nobel Prize) was given to Brian Druker, Nicholas Lydon, and Charles Sawyers for their efforts to develop imatinib for the treatment of CML, "converting a fatal cancer into a manageable chronic condition."[15] This award recognized that the discovery of imatinib represented a major paradigm shift in the field of cancer drug discovery. This breakthrough, which has transformed the lives of patients with CML and changed researchers' thinking about undruggable proteins, started rather simply when Brian Druker was a medical student.

FROM A BCR-ABL INHIBITOR TO A NEW CLASS OF DRUG TARGETS

As a medical student, Brian Druker had little lab experience but developed an interest in cancer chemotherapy. In 1985 he joined the lab of Thomas Roberts, who was working on kinases involved in cancer, at the Dana-Farber Cancer Center. After Druker gained some experience in this field of kinase biology, he initiated a new project in the lab, involving the BCR-ABL kinase.

The Roberts lab, including Druker, began a collaborative effort with Lydon at Ciba-Geigy. Druker developed a key reagent for this collaboration—a specific antibody, known as 4G10, which could efficiently detect the product formed by kinases such as BCR-ABL, allowing for a productive search for chemical inhibitors of this intriguing kinase.

In 1993 Druker moved to Oregon Health and Science University to set up his own laboratory. He continued his productive collaboration with Lydon, testing a series of kinase inhibitors to see if any could be found that killed tumor cells with the BCR-ABL mutation but did not kill normal cells. Within a short period of time they identified imatinib. In the initial clinical trial, within six months they observed that almost all of the patients responded and that there were minimal side effects. Survival for CML patients on imatinib was 89%, versus 50% for those treated with the then-prevailing best options. It was truly a breakthrough, which was unusual in cancer therapy.[16]

Nonetheless, imatinib was not entirely selective for the BCR-ABL kinase over other kinases. This selectivity concern was the original reason that kinases were considered undruggable. In some respects the concern was valid, as it has not been possible to create exquisitely selective kinase inhibitors. However, in the case of imatinib, the drug was selective enough to make it clinically effective.[17] Only a few other kinases are known to be inhibited by imatinib, and it was sufficiently selective to warrant testing in animals and patients.[18]

Indeed, imatinib showed substantial clinical benefit in additional clinical trials and was approved by the FDA (Food and Drug Administration), the United States' regulatory agency charged with determining whether a given drug candidate should be used in patients.[19] Although many patients eventually become resistant to imatinib, its approval is heralded as a defining moment for drug discovery. Imatinib was the first kinase-inhibiting drug approved, and the first time that a specific protein product of a genetic lesion in a tumor was directly targeted by a small molecule drug. For these reasons, there has been a great interest in replicating this success with other types of cancer.[20]

TACKLING THE UNDRUGGABLE PROTEINS

Returning to the concept of druggability, the term *druggable* is an operational definition, based on experiences and hypotheses to date of trying to bind specific kinds of small molecules to specific proteins using specific tests. A challenge with such an empirical definition is that the data supporting the conclusion of undruggability are rarely published—it is difficult or impossible to publish negative results.

This is one of the challenges of the scientific enterprise. Imagine for a moment that you undertake a large-scale screen to test the ability of a million different small molecules to bind to a specific protein. At the end of the experiment, suppose that none of these small molecules has any significant ability to bind to the tested protein. What could you do with this information?

If you tried to publish these findings in any serious scientific journal, they would be rejected. The journal editor would tell you that although the results might be interesting to a specialist such as yourself, most other scientists would not be interested in this information. The problem is that it is difficult to say definitively why the screen failed. For example, perhaps your test was not well designed, and it was incapable of detecting truly effective molecules that do bind to the protein. Or perhaps you used an inappropriate concentration or time point when doing your screen. Maybe the protein you used was not sufficiently pure, or perhaps there was some problem with the equipment.

In other words, there are many reasons to explain negative results, but most of these explanations would only be of interest to the person who carried out the experiment. In order to rule out all of these possible explanations you would have to do an enormous number of additional control experiments to show that the only possible explanation for your failed screen was that the protein of interest is undruggable. While that would be an interesting conclusion and worthy of publication, the required effort would be so massive that it would usually not justify the benefit. This is why negative results are frustrating—they rarely lead to new knowledge, and it is difficult to draw firm conclusions from them.

Much of the discussion of protein druggability is driven by anecdotal accounts of unsuccessful screening efforts, or hypothetical arguments, as with kinases. In the case of kinases, experimental successes eventually changed the perception that this protein family is undruggable. And yet, there is likely something behind the shared anecdotes and failures of many research efforts for other protein families. The challenge is finding the truth behind such stories. What other families of proteins will eventually be found to be druggable as well?

In the coming chapters we will explore some of the latest research that suggests a means by which supposedly undruggable proteins could indeed be targeted with drugs, and what the dramatic impact might be on medicine and disease. The future of medicine is balanced in many ways on this fulcrum of the undruggable proteins. If these proteins are truly resistant to all possible drugs, we will surely see a dwindling supply of new medicines and little improvement against disease. If, on the other hand, researchers find a way to make drugs that bind to these proteins, we will see a bounty of medicines likely causing dramatic improvements against a variety of human ailments. To understand the nature of this critical juncture, we will need to understand the history of drug discovery and where our current drugs have come from.

2

A NEW SCIENCE OF MOLECULES

The challenge of the undruggable proteins is fundamentally about making drugs that can bind to these proteins. In order to understand why it is difficult to make small molecule drugs that interact with putatively undruggable proteins, it is crucial to understand how small molecules came to be used as drugs and how they function mechanistically. The history of small molecule drugs is tightly associated with the emergence of the specific branch of chemistry dealing with carbon-based molecules, known as organic chemistry. This new science of organic chemistry originated with the wonderfully naïve ambitions of a young child in early nineteenth-century Germany.

JUSTUS VON LIEBIG AND THE FIRST SMALL MOLECULE DRUGS

Justus von Liebig played a crucial role in the birth of organic chemistry through the isolation and synthesis of highly purified small molecules and emerged as one of the leading scientists of his generation. Liebig was born in Germany in 1803.[1] He grew up in a home in which chemistry was ubiquitous: his father ran a small business involving oils, paints, and dyes.[2] Liebig's interest in chemistry was likely stimulated by this early exposure to his father's interests and profession. In elementary school Liebig viewed himself as being undistinguished; however, Liebig was two years younger than the other students in his grade, which must have posed a significant social and educational challenge for him.[3]

In his autobiography Liebig recounted an anecdote in which one of his early teachers criticized him in front of all the students in the school, telling Liebig he was the "plague of his teachers and the sorrow of his parents" and asking what he thought he could possibly amount to. Liebig responded by saying he would become a chemist, likely thinking of the humble trade work involving dyes that his father performed. Liebig recounted that upon hearing this response, the whole school broke into an "uncontrollable fit of laughter, for no one at that time had any idea that chemistry was a thing that could be studied."[4]

These events took a toll on Liebig. When he was 14 he dropped out of school.[5] Nonetheless, stimulated by his father's work, he began reading voraciously about chemistry. When he was 16 years old he enrolled as an apothecary's apprentice to learn practical chemistry, trying to prove his classmates wrong. This was his chance to follow in his father's footsteps and to learn a profession that was said to be ridiculous. Unfortunately, his training did not go well. His father related a story that young Liebig blew up part of a building and subsequently had to search for a different means of exploring chemistry; it may be, however, that his father made up this story to cover the fact that he could no longer afford the apprenticeship fee. Whatever the cause, Liebig quit his chemistry training.

After much effort, Liebig eventually returned to school. Remarkably, in time he was able to earn his doctorate and establish himself as one of the most influential chemists of his time.[6] Indeed, the historian of chemistry, James Campbell Brown, said of Liebig, "In some aspects, he may be regarded as the greatest chemist of the nineteenth century, and he is generally recognized as the real founder of organic chemistry."[7] The first drug Liebig discovered was chloral hydrate, which is a sedative and sleep-inducing drug. The challenge in these early days of organic chemistry was to create new molecules, a technique known as organic synthesis because it involves creating (synthesizing) complex, carbon-based (organic) molecules from simpler materials.

In an early example of organic synthesis, Liebig created chloral hydrate by adding chlorine gas to ethanol[8]—the same ethanol molecule that is found in alcoholic beverages. When chloral hydrate is given to patients, it is rapidly converted into trichloroethanol (try-KLOR-oh-eth-ahn-ol), a derivative of ethanol that produces marked sedation.[9] Although the molecular mechanism of action of chloral hydrate is not known in detail, it is thought to function through modulating the excitability of brain cells, although other effects may be involved.[10] Chloral hydrate was one of the first synthetic drugs ever created and marked the start of a revolution in human health—the era of powerful synthetic medicines.

FULMINIC ACID AND ISOMERISM: THE CHALLENGE OF MAKING DEFINED STRUCTURES

In order to make and study small molecule drugs, it is crucial to know exactly what their chemical makeup is. This can be a difficult problem, as Liebig soon discovered. Liebig did pioneering studies on fulminic acid.[11] The science historian Frederick Kurzer has remarked that "there can hardly be another simple compound whose story more consistently reflects the development of organic chemistry over the past two hundred years."[12]

It is worth digressing for a moment here to explain what molecules are. All matter in the universe is made up of the 92 naturally occurring elements that are arranged in the periodic table of the elements, which hangs on the wall of many chemistry classrooms. These elements exist in their simplest form as atoms—small ball-like shapes. Two or more atoms can become connected to each other by means of a chemical bond. Think of the Tinkertoy analogy again. Picture each atom as a ball that can be connected to another ball by a thin rod.

A molecule can be as simple as two atoms connected by a single bond, or as complex as thousands of atoms connected by thousands of bonds, forming extremely complicated three-dimensional shapes. Molecules are also frequently called "compounds" to distinguish them from the simpler elements; compounds (molecules) are created by bringing together different atoms and forming bonds between them.

Fulminic acid is a compound (molecule) that has one hydrogen atom (H), one carbon atom (C), one nitrogen atom (N), and one oxygen atom (O) and, therefore, the molecular formula HCNO. In other words, fulminic acid has four atoms connected by three bonds, arranged in a line.

Acids are so named because they have a hydrogen atom that can be removed in the form of a positively charged atom called an ion. When a hydrogen ion is removed from a molecule, the hydrogen ion enters the solution in which the molecule is dissolved (such as water) and makes the solution acidic. Thus the term *acidity* describes the concentration of hydrogen ions in a solution.

When an acid gives up its hydrogen ion, the remainder of the molecule usually becomes negatively charged. This negatively charged ion can then associate with a positively charged ion, forming a salt. A common example is table salt, which is made up of positively charged sodium ions and negatively charged chloride ions, making the salt sodium chloride (NaCl).

Liebig was interested in understanding why salts of fulminic acid, such as mercury or silver salts, are highly explosive.[13] Mercury fulminate, the mercury salt of fulminic acid, has been used in blasting caps as a trigger to ignite other explosives and was exploited by Alfred Nobel to create dynamite. After accumulating wealth from this discovery, Nobel used this money to establish the Nobel Prizes.[14] Silver fulminate, sometimes used in noisemakers and firecrackers, is even more explosive.[15]

At the same time that Liebig was studying these fulminate salts, the great German chemist Friedrich Wohler synthesized and studied isocyanic acid.[16] Curiously, this molecule, like fulminic acid, was composed of one atom each of hydrogen, carbon, nitrogen, and oxygen. It has the same molecular formula as fulminic acid (HCNO). However, the salts of isocyanic acid, such as silver isocyanate, are not explosive.[17]

This was a profound mystery that threw into question the emerging principles of this science of organic chemistry. How could two different molecules have the same chemical composition but dramatically different properties? It seemed to be an impossibility. Wohler and Liebig puzzled over this dilemma. Molecules made up of the same elements should have the properties—they should be identical.

Liebig was excessively bold, to the point of being rude, in criticizing others; he pointedly insisted that Wohler had erred in his analysis, implying that Wohler was incompetent. Wohler persisted in defending his own analysis. An impasse was reached. In other spheres of life this might have led to a duel or to a lifelong feud over competing ideas. The beauty of science is that, at the end of the day, there are always data.

Ultimately, working together, Liebig and Wohler determined that they were both correct. Liebig was gracious enough to apologize for calling Wohler incompetent. They proceeded to become close friends and longtime collaborators. In this dispute, Wohler and Liebig had discovered the first example of *isomers*, that is, two different molecules with the same formula but different connectivity. They found that fulminic acid has the connectivity H-C-N-O, in which the oxygen atom is connected to the nitrogen atom, whereas isocyanic acid has the connectivity H-N-C-O, in which the oxygen atom is attached to the carbon atom. This turned out to be a fundamentally important concept for molecules. The properties of a molecule are determined not just by the type of elements that make up the molecule, but also by the precise connectivity of these elements.

From these findings, it became clear that when creating new molecules, it is crucial to define their exact composition, not only in terms of what atoms make up the molecules, but also in terms of how these atoms are connected to each other in space. With this understanding, it would become possible to create complex and powerful medicines.

SYNTHESIS OF BIOLOGICAL MOLECULES

As the basic principles governing the chemistry of small molecules were being elucidated, two related approaches for generating biologically active molecules were pursued: synthesis and extraction. Biologically active molecules are the source of future drugs, so figuring out how to obtain them was a crucial challenge.

Wohler catapulted the synthesis approach forward when he accidentally synthesized urea in 1828. It was recognized at the time that organic compounds (i.e., those containing carbon) were derived from living organisms, usually plants or animals. Much of the public, and many scientific researchers, such as Jons Jakob Berzelius (a leading chemist of the time), suspected that these organic molecules contained a vital force within them; by this logic, they could not be created from inorganic materials.[18] According to this view, there was a hard line demarcating the molecules of life from the molecules of inanimate materials.

It was therefore a great surprise when Wohler added ammonium sulfate to potassium cyanate, both inorganic molecules found outside living systems, and found that instead of creating ammonium cyanate as he intended, he had made something that was impossible—urea, an organic compound found in the urine of living, breathing animals. This would be the equivalent today of mixing two plastic materials together and finding a small animal suddenly appear out of the mixture. For many, this was truly incomprehensible.

Wohler's discovery was the beginning of the end of the notion that a vital force was present in compounds from living systems—if you could make molecules normally found in live animals by using inanimate materials, then there was no special force present in molecules derived from living matter. One after another over the succeeding years, it was shown that all compounds found in nature could be synthesized in the laboratory from simpler materials found outside of living systems. In other words, molecules are molecules and it doesn't matter whether they are made in nature or in the laboratory. The same molecules will have the same properties, irrespective of how they are created.

In 2010 J. Craig Venter and colleagues performed the modern equivalent of Wohler's feat: they succeeded in creating a type of synthetic cell. They did this by synthesizing a huge, artificial piece of DNA and using it to replace the normal DNA found in a bacterial cell. DNA is found inside cells and contains the instructions for operating the cell and for creating new cells. Thus the instructions for this artificial cell were completely designed and created by Venter and colleagues, rather than having been produced in the normal fashion by nature.[19] Once again it was found that a molecule is a molecule, whether it is made in a laboratory or by a cell.

EXTRACTING AND PURIFYING COMPOUNDS

At the same time that synthetic chemistry was emerging as a means to create medicines, extraction methods were being perfected to purify molecules from natural sources. These extracted natural products could have profoundly beneficial or malignant properties, depending on the molecule. The goal of such purifications was to find a way to separate one compound out of a mixture present in a natural source, such as a bacterial culture broth or a plant. For example, when you grind up coffee beans and add hot water, you extract chemicals present in the coffee beans into water, forming coffee. This coffee bean extraction procedure retrieves multiple chemicals from the coffee beans into your coffee beverage, and a further purification effort would be needed to separate caffeine from the other flavor molecules present, for example.

Although herbalists had been extracting molecules from plants for centuries, the concentration of active compounds present in these extracts was often variable, depending on the weather, age of the plant, the season, the length of the extraction, and other parameters. Since each extract contained dozens or hundreds of different molecules, it was impossible to create a reliably active preparation that would have the same properties every time.[20]

The solution to this dilemma was to identify a single ingredient in a given extract and then standardize the concentration of this purified compound. In this way it would be possible to create a therapy that would have the same activity every time it was administered to patients and to reliably test compounds for both benefit (i.e., efficacy) and side effects (i.e., toxicity). Charles Louis Cadet de Gassicourt was a leading proponent of this viewpoint.[21] He argued that pure substances should be used in place of complex extracts.

The first of these purified natural compounds was morphine. Friedrich Serturner purified morphine from the poppy plant in 1817.[22] Serturner's father was an engineer and surveyor for the local royalty, and Friedrich originally planned to follow in his footsteps; however, his father's death forced him to take a job in the field of pharmacy.[23]

Since almost all organic (carbon-based) molecules are more soluble in hot water than in cold water, Serturner added hot water to opium to separate the morphine from other materials and then added ammonia to cause the morphine to fall out of solution (i.e., to form a precipitate).[24] One of the fundamental properties of all molecules is their solubility in different liquids (called solvents). Synthetic chemistry, the branch of chemistry that deals with making molecules, relies upon putting two compounds into a solution together to cause them to react and form a new molecule. Moreover, purifying a compound from a plant or other natural source requires taking advantage of differential solubility of molecules in different solvents. Testing a compound in animals or humans requires that the compound be reasonably soluble in the bloodstream, as well as in other body compartments. A molecule's ability to dissolve in different solvents is thus a fundamentally important property.

One property that is critical in predicting the ability of a molecule to dissolve in a solvent is the hydrophobicity of the molecule and the solvent. The term *hydrophobic* is derived from *hydro*, for water, and *phobic*, for fearing. Thus, hydrophobic molecules and solvents are water fearing. In contrast, hydrophilic molecules and solvents are water loving. A hydrophobic molecule, like the free base form of morphine, does not dissolve well in water. Instead, it dissolves in solvents that are more hydrophobic, like ethanol. A hydrophilic molecule like table salt (NaCl) dissolves well in water but does not dissolve well in hydrophobic, greasy solvents.

How do you tell whether a molecule is hydrophilic or hydrophobic? The most accurate way is to measure this property experimentally in the laboratory, although there are also computer-based prediction methods.[25] If you have a molecule of interest and are curious to know how hydrophobic or hydrophilic it is, you can add the molecule to a mixture of water and octanol, which is an oily hydrophobic solvent that will not mix with water. These solvents form two layers in a test tube that are separated from each other, like oil and vinegar. By measuring how the test molecule partitions between the water and the octanol layers, you can quantify exactly how water loving or water fearing it is. This is known as the partitioning, or P value, for the molecule. The P value is critical for determining how the molecule will behave when injected into people or animals and how the molecule should be handled in the laboratory.

An esteemed medicinal chemist, Chris Lipinski, formerly from Pfizer, published an analysis of the properties of molecules that make for good drugs.[26] One part of this rule specifies that the P value for the drug candidate should be between 1 and 100,000. If a drug is too hydrophilic ($P < 1$), then it won't penetrate across the greasy membrane of cells. On the other hand, if a drug is too hydrophobic ($P > 100{,}000$), then the molecule will be poorly soluble in the bloodstream and in the interior of cells, and it will have a hard time getting where it needs to be to exert a beneficial effect. This is one of the ways in which drug developers now try to prioritize future drug molecules—by picking molecules with suitable P values.[27]

Returning to the purification of morphine by Serturner, he found that morphine is a basic molecule. In acidic solutions it picks up a proton (hydrogen ion) and becomes charged and therefore has high water solubility (low P), but in basic solutions it is not charged, so it has poor water solubility (high P). Serturner used this property to purify one of the first compounds from natural sources. This molecule, morphine, represented a major scientific milestone. It opened up a vast realm of possibilities, in which researchers would troll the depths of the oceans, scour the plants of the Amazon jungles, and harvest bacteria in soil, looking for powerful new medicines to extract from natural sources. Like many scientific discoveries, however, there was an unforeseen consequence of this purification effort. In time, morphine became widely used and abused; during the U.S. Civil War, morphine was widely adopted for pain relief by soldiers, but many became addicted.[28] Then, in an advance of questionable value, the pharmaceutical company Bayer synthesized a diacetyl (dai-ah-seh-TEEL) derivative of morphine, which they named heroin.[29] As with morphine, heroin use also became widespread, along with concomitant addiction. It was eventually banned, leading to a near cessation of its official use.

Thus two powerful technologies were developed for obtaining potential drug molecules: synthesis in the lab and purification from natural sources. These two methods have continued to provide biologically active compounds, and potential drugs, up to the present day.

COMBINING EXTRACTION AND SYNTHESIS TO MAKE FEVER-REDUCING DRUGS

In many cases, extraction and synthesis methods have been combined to make derivatives of molecules found in nature. An example of this type of combined approach is the discovery of fever-reducing drugs.

An active component of Willow tree extract is salicin.[30] Raffaele Piria, an Italian chemist, prepared salicylic acid from salicin, which was advantageous because salicylic acid was easier to modify into other molecules compared to salicin.[31] Salicylic acid, however, caused gut toxicity, and so was of limited value.[32] Charles Frédéric Gerhardt, a French chemist, then synthesized a key derivative named acetylsalicylic acid in 1853.[33] Refined methods for synthesizing acetylsalicylic acid were developed by two chemists, Johann Kraut in 1869 and Hermann Kolbe in 1874, but then, unfortunately, this molecule languished, with little interest in its medicinal properties.[34]

At the end of the nineteenth century, many researchers began to pursue medicines that could reduce fever, with the rationale that fever could be damaging to the body. These so-called anti-pyretics (fever-reducing molecules) were aggressively sought. The nascent pharmaceutical company Bayer introduced acetylsalicylic acid as an anti-pyretic in 1899, coining the name "aspirin."[35] This fever-reducing drug quickly became popular around the world. It was the first of the non-steroidal anti-inflammatory drugs used (NSAIDs). Subsequently, other NSAIDs were introduced, such as ibuprofen (now sold as Advil). Such drugs represent a huge class of beneficial medicines that are widely used around the globe.[36]

Stepping back a moment to look at how these pioneering molecules were discovered during the emergence of this era of the first drugs, we can see a theme. Morphine is a naturally occurring compound that was purified by extraction and subsequently modified synthetically to make many different pain-relieving drugs. Chloral hydrate was prepared by a simple reaction between two readily available compounds. Salicylic acid and aspirin were prepared by simple synthetic modifications of a natural compound, salicin.

Thus, if we ask where these drug molecules came from in these early days, we find that they were either purified from natural sources or prepared using very simple chemical reactions from commonly available materials or natural products. In these early days no one had any idea what proteins were bound by these drugs in cells. This certainly didn't enter into the drug discovery process. Instead, these researchers simply began testing natural compounds or simple synthetic derivatives in animals, and in some cases found strikingly active molecules. It wasn't a very efficient process, but on the other hand, there weren't many molecules available to test. It was painstaking work to purify a molecule from natural sources, and laborious to run even simple chemical reactions and purify and characterize the product of a reaction, so there were only a small number of drug candidates that could be tested. Over time, methods for creating and extracting compounds became more robust; in the modern era, nearly any compound can be made in the lab or extracted from natural sources.

However, even with these developments, a major challenge remained. What proteins do drug molecules bind to? Initially there was little focus on this question, as the knowledge of biology was not sufficient to determine this level of molecular detail. Eventually, though, this became an increasingly crucial question. Indeed, it is central in the quest to target the undruggable proteins.

FINDING THE TARGET PROTEIN FOR MORPHINE IN CELLS

How do powerful analgesics like morphine and heroin function, and what proteins do they interact with in our bodies and cells to exert such a powerful effect in relieving pain? An answer was found for morphine when Solomon Snyder, a professor at Johns Hopkins University, and his graduate student, Candace Pert, addressed this difficult question.[37]

First they created a morphine-like molecule that functioned as a molecular tag, to assist them in identifying the protein that interacts with morphine. They synthesized a derivative of naloxone, which is chemically similar to morphine, and attached a radioactive label to the molecule. This "radiolabeled" molecule contained tritium, which is a small, radioactive element that gives off an easily detectable signal. Once naloxone had been labeled with tritium, it was possible to follow the location of the tritium-naloxone in a test tube by looking for the tritium label, which is like a bright beacon pointing out where the molecule is located.

Pert and Snyder then ground up brains of rats in order to purify the protein that binds to morphine-like molecules called opioids.

They added tritium-naloxone to the mixture of brain proteins. To determine how much of the tritium-naloxone had bound to proteins, they captured the proteins on a material called a filter, washed the filter to remove excess tritium, and pulled the proteins off of the filter with a detergent. They then measured the amount of tritium present in this detergent-extracted sample and found a large amount. This suggested that the tritium-labeled naloxone could bind to proteins from rat brain.

However, one crucial skill of researchers is their ability to think of the best control experiments—the ones that can potentially invalidate their own conclusions. As researchers design better control experiments, they become more confident in their interpretation of their data. If one of the control experiments refutes their interpretation, they save themselves from drawing a conclusion that is fundamentally incorrect.

In the case of naloxone-binding activity, Pert and Snyder realized that the tritium-naloxone could bind to proteins through numerous mechanisms. Some small molecules are intrinsically sticky, and naloxone might be one such molecule. Some proteins are also intrinsically sticky and can allow many different compounds to weakly bind to them. This kind of weak binding between a small molecule and a protein is called nonspecific binding. If you picture a protein as a crystal pitcher, then specific binding would represent a lid that snugly fits into the top of the pitcher, perfectly matching the shape. Nonspecific binding would represent a piece of gum stuck to the side of the pitcher. The hallmark of specific binding is that it can be prevented by a similar molecule binding in the same position on the protein.

In the case of the crystal pitcher, if a lid were already in place, it wouldn't be possible to place a second lid onto the pitcher. The lids compete with each other for binding to the pitcher. The gum, on the other hand, is nonspecific. If you have a piece of gum stuck to the side of the pitcher, you are still able to stick a second, third, and fourth piece of gum onto the pitcher. In other words, specific binding is found when there is a single binding site on the protein. Nonspecific binding occurs where there are a large number of binding sites.

To examine specific binding, Pert and Snyder tested whether morphine could prevent the labeling of rat brain proteins by tritium-naloxone. If morphine and tritium-naloxone bind to the same protein, which we would expect given that they are very similar molecules, then the presence of copious amounts of morphine should saturate the protein. There would no longer be any place for the labeled naloxone to bind. In the pitcher-lid analogy, if the morphine lid is already in place, it is not possible to put the naloxone lid in the same position. This experiment revealed that morphine and naloxone bind competitively to the same protein in a rat brain, named the opioid receptor. In fact, naloxone is used to treat overdoses of morphine or other opioids. Naloxone reverses the effect of morphine overdose—morphine is powerless to affect the opioid receptor when naloxone is present.

Shortly after these experiments, Hughes and Kosterlitz, as well as Sol Snyder's group, isolated a peptide normally present in animal brains that competes for binding to the same protein receptor.[38] Peptides are short pieces of protein that are built up from the same repeating amino acid units from which proteins are built. A number of similar peptides were subsequently isolated that also bind to this opioid receptor. These peptides were named *endorphins*, for endogenous morphines, because they naturally exist in the brain of animals to activate these pain-relieving receptor proteins. Thus the study of morphine, a plant-derived natural product that fascinated researchers for 100 years, led to the discovery of functionally similar molecules that normally reside in human brains. Morphine hijacks this natural system and overstimulates it, causing an overwhelming sense of pain relief and euphoria.

THE EMERGENCE OF MOLECULAR MEDICINE

In the early years of organic chemistry, in the era when morphine, aspirin, and chloral hydrate were discovered, the focus was on making drugs, and there was little concern with how they acted on a

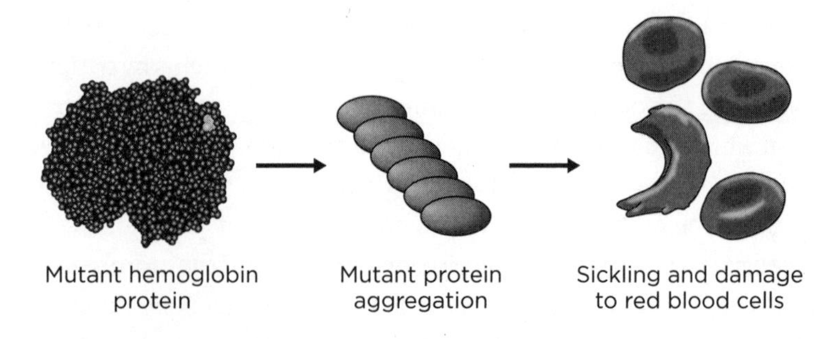

Figure 2.1 Effects of sickle-cell mutation. Hemoglobin is a protein that carries oxygen in red blood cells and that is mutated in patients with sickle-cell disease. The mutation causes the mutant hemoglobin protein to aggregate, changing the shape of red blood cells and leading to disease symptoms.

molecular level. In the mid-twentieth century, with the slow emergence of a molecular understanding of disease and drug action, the approach to drug discovery began to change, and the foundation was laid for defining the problem of the undruggable proteins. This shift can be traced to 1956, when Vernon Ingram found that sickle-cell anemia was caused by the mutation of a single amino acid in the hemoglobin protein (see Figure 2.1).[39]

Proteins are strings of amino acids stitched together through peptide (amide) bonds like beads on a string. It was unexpected that substituting just one of these amino acids for a slightly different one could cause such a dramatic change in one protein such that a disease would result. Ingram is often called the "father of molecular medicine" because this was a turning point in terms of understanding disease from the perspective of specific molecules going awry.

Ingram joined the faculty at Massachusetts Institute of Technology (MIT) in 1958 for a one-year visit and ended up having a long and productive career there. In 2002 he was elected to the U.S. National Academy of Sciences. I have to admit that I was unaware of Vernon's seminal contributions in 2001, when he called me on the phone to ask if I wanted to collaborate with him. I had started as an independent Fellow at the Whitehead Institute at MIT in the fall of 1999. One of the first things I did when I arrived to set up my new lab was to assemble a collection of 2,000 biologically active molecules.[40] My goal was to assemble thousands of compounds that were already known to be biologically active, and to determine whether any of these compounds were active in new disease models.

Vernon heard about my research efforts, I must assume, because he called and expressed interest in testing my collection of molecules in his cellular model of Alzheimer's disease. Alzheimer's is a common form of dementia, originally identified by Alois Alzheimer in 1906, that involves memory loss, confusion, and eventually loss of bodily functions. One of the characteristic features of the disease is the presence of insoluble protein aggregates in the brains of patients. These insoluble plaques are formed from a fragment of a protein named beta amyloid.

The mechanism connecting beta amyloid aggregates to memory loss is not known, but Vernon was interested in answering this question. He proposed that we test my collection of 2,000 known, biologically active compounds to see if any of them could reverse the effects of beta amyloid on neurons. He had found that when he added aggregated beta amyloid to neuronal cells, the membrane properties of the cells were altered, and he wanted to know if any of the compounds could reverse this effect of beta amyloid.

Vernon and his research assistant Barbara Blanchard worked with me to test all 2,000 of our compounds in his assay. We found several that were effective at reversing the effects of beta amyloid in this assay, and we soon published a paper on this finding.[41] Vernon and Barbara went on to define how one of these compounds, named DAPH (4,5-dianilinophthalimide), functioned.[42] They found that it bound directly to aggregated beta amyloid and prevented it from interacting with neuronal cells. Vernon was interested in trying to optimize this compound and turn it into a drug. It hasn't yet made it that far, however, as drug development is a long and tortuous road.

I remember with great fondness our collaborative discussions. Sitting in Vernon's office, I was surrounded by a thick carpet and floor-to-ceiling bookshelves lined with dense tomes. When I visited the offices of most other professors, there was a constant roar of activity—the phone would be ringing, the fax beeping, the e-mail alerts coming in one every ten seconds, and students and colleagues would be popping their heads in to ask questions. In most offices, there is constant motion, but little time to think. That is why I so loved my visits to Vernon. There was a tranquility there, like a Japanese book garden for scholars. We would sit and contemplate data without interruption, having time to think deeply. I felt Vernon was just the sort of contemplative scholar I hoped to become one day, and I hoped to make the space between life's daily events for this kind of thoughtful scholarship.

When I moved to Columbia University in 2004, our collaboration gradually ended, in part because of the distance and in part because my role was winding down on the project. However, we still communicated from time to time. Vernon was kind enough to write several letters of recommendation for me, and I offered him advice about chemical libraries and how to screen them as he sought to test additional compounds. He was always modest, gracious, and supportive.

It was with shock that I read a news article forwarded to me in the fall of 2006. It said that Vernon slipped and fell and ultimately died from his injuries on August 17, 2006. He was 82 years old. I was distraught over this news, as I had spoken to Vernon not long before. It highlighted for me the transient nature of life and of our interactions with friends, mentors, and colleagues in science. We become accustomed to building relationships that deepen year after year, and we forget that life is precious and ultimately finite. It was with great sadness I realized I would never sit and contemplate data with Vernon again.

DESIGNING TARGETED DRUGS

The era of molecular medicine, ushered in by Vernon, brought with it an appreciation that discovering which proteins are targeted by small molecule drugs is crucial to understanding their mechanism of action. Much later it became clear that some proteins were easy to target with drugs, while others were much more challenging to target and appear undruggable.

The emergence of drugs that interact with specific proteins actually pre-dated Vernon's discovery of the sickle-cell mutation. This approach began with the discovery of sulfanilamide (SUL-fah-NIL-ah-mayd) as an antibacterial agent by Gerhard Domagk and Therese Trefouel in 1935.[43] Sulfanilamide was found to have an unexpected diuretic effect, in which the rate of urination was increased, causing increased excretion of water.[44] This was of interest, because diuretics were being explored as a means of treating hypertension, or high blood pressure.

In 1940 Thaddeus Mann and David Keilin hypothesized that the reported diuretic effects of sulfanilamide could be due to inhibiting a protein named carbonic anhydrase, which was already known to be important in excretion through the kidneys.[45] Mann was a biochemist born in Austria-Hungary who studied medicine in his home country before moving to Cambridge in 1935 and then turning his attention to sulfanilamide. David Keilin was an entomologist, born in 1887 in Moscow. He eventually wound up working with Mann in Cambridge at the Molteno Institute, as Keilin's work took on a biochemical flavor. More potent carbonic anhydrase inhibitors were synthesized, leading to a major class of drugs that are used effectively to treat hypertension today. This approach to drug discovery represented a major shift, in that a specific protein, carbonic anhydrase, was targeted with small molecules. Gradually, the entire pharmaceutical industry shifted to use this targeted approach. In fact, it is challenging to develop a drug today without knowing its molecular target.

This theme of targeted drugs has become increasingly effective, as can be seen in the subsequent discovery of beta-blockers for treating hypertension. George Oliver and Edward Schafer, at University College in London, discovered that if they injected an organic extract of an adrenal gland into animals, the animals' blood pressure would rise.[46] The active molecule, epinephrine (also known as adrenaline), was discovered in 1901 by Jokichi Takamine and then synthesized in the laboratory by Friedrich Stolz in 1904.[47] The molecular target of epinephrine remained elusive until the American scientist Raymond Ahlquist found in 1948 that there are two types of protein receptors for epinephrine: alpha receptors and beta receptors.[48] In 1963 James Black discovered the powerful compound propanolol, which could bind to the beta receptors and prevent epinephrine from turning them on.[49] Such a compound, which blocks activation of a receptor is called an *antagonist*, or a blocker. Propanolol is known as a beta-blocker, because it blocks activation of the beta receptors for epinephrine. Propanolol prevents the rise in blood pressure that would occur when epinephrine is released, which decreases the force of heart contractions and stress on heart muscle. This discovery was recognized with the Nobel Prize in 1988. Propanolol and related beta-blockers are still used today for effectively treating hypertension. Thus a valuable targeted drug was successfully created.

STRUCTURE-BASED DRUG DISCOVERY

As one drug target after another was discovered and validated, the pharmaceutical industry became increasingly profitable and increasingly focused on elucidating these proteins that could serve as drug targets in order to understand the detailed mechanism of action of drugs. The promise of this approach was that it should be easier to design and optimize drugs if you could determine which proteins they need to interact with. Moreover, there was an intellectual satisfaction that came with knowing the precise mechanistic details of how a drug exerted its therapeutic effects that was thrilling to molecular scientists.

In these early research efforts the drug came first, followed by the drug target. Before long, drug discovery researchers began to select proteins first, and then tried to design drugs that could bind tightly to the designated protein. This approach represented the final triumph of molecular biology, as the hope was that by understanding disease mechanisms in terms of which proteins go awry, it should be possible to design drugs a priori that modify diseases by binding to these proteins. In such cases, drugs are designed using information about the three-dimensional shape of the designated protein. This three-dimensional shape is called the structure of the protein.

One of the first examples of this structure-based drug design came in the search for drugs to treat acquired immunodeficiency syndrome (AIDS). In 1981 a striking rise in pneumonia infections began among homosexual men.[50] This marked the start of a viral outbreak that was termed AIDS.[51] More than 60 million cases have been diagnosed around the world.[52] In 2008 a Nobel Prize was awarded to Francoise Barre-Sinoussi and Luc Montagnier for the discovery of the virus that causes AIDS. Francoise Barre-Sinoussi, Luc Montagnier, and Robert Gallo identified this virus, which came to be called the human immunodeficiency virus (HIV).

The first drugs developed to treat HIV infection were similar to the building blocks of DNA. These drugs prevented the virus from replicating its genetic instructions. These drugs were effective, but fairly toxic to normal cells, too, since they could also inhibit the replication of normal cellular DNA.[53]

A second strategy for treating HIV infection was to inhibit the HIV protease, which is a protein-cleaving enzyme that is needed for HIV to function in cells. The three-dimensional structure of HIV protease was determined, and small molecules that could bind to the protease and block its activity were identified. The first of these was approved in 1995.[54] This approval heralded the ascendency of structure-based drug design in the pharmaceutical industry. It had become possible to design drugs that bind to a specific protein, and this had a dramatic impact on disease. HIV protease inhibitors, when used in combination with modified DNA building blocks drugs, were able to control HIV infections for the first time, turning HIV infection from a fatal disease to one that was manageable.

The challenge in twenty-first-century drug discovery has now become selecting suitable drug targets. Which proteins should researchers target in order to find new medicines that can cure intractable diseases? In order to answer this question, we must examine in detail the kinds of proteins that are attacked by specific drugs. To do this, we will look at the origin of cancer chemotherapy, and whether there is any hope of developing drugs for the many different types of cancer that are currently incurable.

3

THE BIRTH OF THE FIRST CANCER DRUGS

In the modern era, discovering effective anti-cancer drugs requires identifying suitable cancer-regulating proteins that drugs can bind to. Which proteins have been chemically tractable, leading to effective cancer drugs, and which currently intractable proteins might be targets of drugs in the future? To answer this question let us look back at the origins of today's cancer drugs.

THE DISCOVERY OF SULFUR MUSTARD

In 1886 the German chemist Viktor Meyer for the first time synthesized in pure form 1,5-dichloro-3-thiapentane (die-CHLOR-oh THIY-ah-PENT-ane)—an oily fluid with a subtle, sweet smell.[1] Little did he know that this simple molecule would both change the nature of warfare and usher in the first drugs for treating cancer.

Meyer was a Jewish German and the son of cotton pattern printer who made a living by printing designs on cotton fabrics. Meyer's parents were not active in science, and Meyer grew up loving poetry and acting. It was the influence of his older brother that exposed him to chemistry, which he would grow to love even more than poetry.

Meyer received his doctoral degree at the age of 19, leading to a meteoric rise in the world of chemistry.

Meyer found that his newly synthesized molecule caused large blisters to appear wherever it contacted skin, primarily on his research assistant.[2] This was before the days of OSHA (the Occupational Safety and Health Administration within the U.S. Department of Labor), which protects the safety of workers through strict regulations. Meyer initially thought his assistant might have manufactured these symptoms, so he had the chemical tested on rabbits at a local medical school, where it was found to be lethal at moderate doses. He discontinued study of this poisonous substance, which nonetheless was intriguing to him because of its striking physiological effects.[3]

Several years later, Hans T. Clarke refined the synthesis of this compound and confirmed Meyer's observations of its toxic properties.[4] Clarke broke a container holding the compound, which caused a severe injury to his leg, causing him to be hospitalized for two months.[5] Clarke was another leading chemist of his day, born in Harrow, England to an American father and German mother in 1887. Unlike Meyer, Clarke was exposed to science and medicine from an early age. Clarke studied chemistry and physiology at University College in London before moving to Berlin to work in the laboratory of the great German chemist Emil Fischer. Clarke later founded the organic chemistry division of Kodak and was its first employee.

He eventually moved to Columbia University as a professor in the Biochemistry Department at the Columbia University Medical Center.[6] There, Clarke trained Konrad Bloch, who would go on to win the Nobel Prize in Physiology or Medicine in 1964 for his work on the mechanism of cholesterol and fatty acid metabolism.[7] Jie Jack Li, a science writer, has suggested that Clarke was inclined to train Bloch because of Bloch's proficiency with the cello and Clarke's love of music; whatever the reason, Bloch completed his PhD in just a year and a half, so he turned out to be an efficient student.[8]

Clarke came to believe later in life that the report of his early lab accident in Berlin to the German Chemical Society caused the Germans to develop his powerfully toxic compound into a chemical warfare agent.[9] The compound was first deployed on July 12, 1917, during World War I, when Germans fired upon English troops with shells containing 1,5-dichloro-3-thiapentane, or what has become known as sulfur mustard, because of its yellowish brown color.[10] It has a garlic- and mustard-like odor, comes in the form of an oily liquid, and is highly toxic as both a gas and a liquid. Sulfur mustard penetrated through clothing and accumulated on the ground, causing more severe damage than a purely airborne chemical agent would have.

The Germans had experimented with other poisonous substances during World War I, such as chlorine gas,[11] but sulfur mustard was the most effective, as it could be loaded into shells and delivered as a projectile to enemy soldiers, in contrast to chlorine gas clouds that could blow back onto the Germans.[12] Those exposed to sulfur mustard underwent progressively more devastating responses. First, they experienced acute damage to the eyes, because the corneal epithelial cells rapidly divide. Their eyelids showed signs of severe burns, and large lesions were produced in their corneas, causing blindness. Pain soon became evident in the nose and sinuses, along with sneezing, a hacking cough, and a sore throat. Soon, burning skin blisters appeared under their arms and on their necks, faces, and genitals,[13] followed by airway obstruction and respiratory failure, frequently leading to a painful death.[14] Thousands were killed after their first exposure.[15] Sulfur mustard has been used in more than 10 conflicts over the years; the number of casualties caused by sulfur mustard outnumbers those caused by all other chemical agents combined. With the resurgence of global terrorism in the last decade, there is now increasing concern about the use of this agent by terrorist groups.[16] Clarke and Meyer unleashed a terribly effective chemical upon the world. Unbeknown to them, they also laid the foundation for the first cancer drugs.

FROM CHEMICAL WARFARE TO A CANCER DRUG

Interest in sulfur mustard waned after World War I but was resurgent during World War II. Related compounds were created, and this class of agents divided into the sulfur mustards and the nitrogen mustards. Mechanistic studies revealed that the lethality of mustard agents to cells was dependent on the rate of cell proliferation. Initially, this research was classified, as it dealt with improving chemical warfare agents. In 1946, however, Major Alfred Gilman and First Lieutenant Frederick Philips published an article in the journal *Science* describing preliminary analyses of how these compounds affected cells; they suggested that in addition to their use in warfare, these agents might have applicability in the treatment of some cancers.[17]

To understand the basis for the profound toxicity of these mustard agents, we need to examine their chemical structures. The key feature of the original sulfur mustard is that it has a sulfur atom, followed by two carbons and a chloride leaving group: S-C-C-Cl. Some atoms, such as chlorine, are known as good "leaving groups"—they like to break away from carbon atoms and go off on their own—in this case forming a chloride ion.

Simply having a good leaving group such as chloride, however, doesn't make a molecule into a mustard agent, with its associated toxicity. For example, the molecule ethyl chloride, with the functionality C-C-Cl also has a good chloride leaving group, but it is not particularly toxic. The severe toxicity of sulfur mustard involves the unique location of the leaving group—two atoms away from the sulfur atom.

Sulfur (as well as nitrogen) is the opposite of a good leaving group; it is a good nucleophile (NOOK-lee-oh-fayl), meaning that it likes to attack other atoms. Thus the sulfur atom can flip around and cause the chloride to dissociate (see Figure 3.1). This creates a small ring made up of one sulfur and two carbon atoms. The ring is tightly wound and strained, creating a molecule that will react rapidly with other molecules. It is like a spring-loaded trap, waiting to be opened.

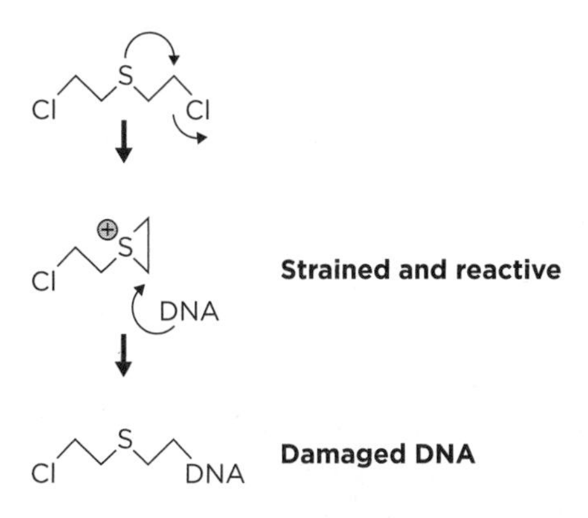

Figure 3.1 The mechanism of sulfur mustards. Sulfur mustards have a central sulfur atom that can donate its electrons to a nearby carbon atom, bearing a chloride leaving group. Loss of the chloride group causes a strained ring to form, which can be readily attacked by DNA, leading to damage of the DNA.

It can react with proteins or DNA in cells, thereby damaging these cellular constituents.

This is the basis for the striking toxicity of mustard agents. Gilman and Philips noted that treatment of the fruit fly *Drosophila melanogaster* with mustard agents caused a high rate of mutation resulting in lethality, something that had previously been observed only with powerful radiation.[18] Thus mustard agents were remarkably effective, having the ability to alter the basic DNA instructions of cells.

Treatment of animals or humans with mustard agents revealed suppression of blood-forming cells.[19] Gilman and Philips suggested on the basis of this observation that mustard agents might, strange as it may seem, have a beneficial application in treating diseases, such as leukemia, caused by excess proliferation of hematopoietic (hem-AT-oh-poy-et-ik), or blood forming, cells. This was an inversion of the nature of these agents—they were transformed from being a nefarious chemical weapon designed to cause maximum casualties into a potential life-saving medicine against an incurable disease.

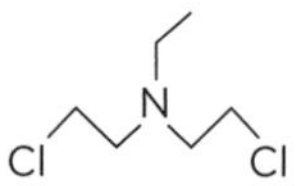

Cl N Cl

Figure 3.2 Example of a nitrogen mustard. Nitrogen mustards are similar to sulfur mustards, but the central sulfur atom is replaced by a nitrogen atom.

Initial experiments in mice showed that lymphosarcomas could be cured with nitrogen mustards (see Figure 3.2), although there was a fine line between the therapeutic dose and extreme toxicity to normal tissues.[20] This would become a long-running theme in the world of cancer treatments. In 1943 Gilman and several colleagues treated the first patients with mustard agents.[21] Out of six patients with terminal disease, two had their tumors rapidly dissolve, encouraging further testing in additional patients.[22] Sixty-seven patients with a variety of tumors were then treated by this group of researchers. They found that Hodgkin's lymphoma was most susceptible to mustard agents, whereas other tumors of the blood system were less susceptible. They noted that side effects, such as reduction in numbers of normal blood cells, while undesirable, were tolerable.

A typical patient studied was L. W., a 33-year-old housewife.[23] She was diagnosed in 1941 with Hodgkin's disease, a type of lymphoma. Her father had died of Hodgkin's lymphoma, suggesting a possible genetic basis for her disease. Her lymphoma spread rapidly throughout her body. Initially, radiation therapy was effective, but over time it became less effective. Over the next summer, L. W. developed shortness of breath and a cough, as well as fluid around her lungs. She became increasingly weak as the disease progressed. By the spring of 1943 she was forced to remain in bed and no longer responded to radiation. By December her condition worsened and she was admitted to a hospital; she was considered extremely fragile. She had a fever, a shallow, dry cough, and gasped for breath continually. Her face, upper body, and right arm were swollen, and large masses protruded from her neck and from under both arms. Her right arm had become paralyzed.

In this troubled state, L. W. was treated with a nitrogen mustard, via syringe injection, every other day for two weeks. She reported feeling much better, her fever subsided, the lumps (actually swollen lymph nodes) shrunk dramatically, and the swelling of much of her body was eliminated. Moreover, minimal side effects were seen with the drug.

There was restrained excitement for both the patient and the physician at seeing this profoundly positive improvement in a patient with a devastating lymphoma—how remarkable that the effect was produced by a chemical warfare agent. Ultimately, sadly, the effect was short lived. The improvement continued for four weeks, and then one day L. W.'s lungs filled with fluid suddenly, causing her rapid death. Despite this reversal, L. W.'s treatment had launched the era of cancer drug therapy.[24]

CYCLOPHOSPHAMIDE IS BORN

Shortly after this pioneering work by Gilman and colleagues, Norbert Brock, a young oncology researcher, completed his training in pharmacology and medicine in Berlin.[25] In 1949 he was appointed head of the Pharmacological Department at ASTA Werke AG, a medium-sized pharmaceutical company now known as Baxter Oncology.

Others had proposed an approach to anti-cancer drug discovery involving the creation of a pro-drug that would be converted by tumor cells into a highly reactive molecule. Brock and his colleagues at ASTA attempted to implement this idea in the context of nitrogen mustards. Their strategy was to reduce the general toxicity and side effects of the mustard agents by masking them in an inactive state. If mustard analogs could be designed in such a way that they were only activated in tumor cells, they might specifically target these tumor cells over normal cells. The N-C-C-Cl fragment of the mustard molecule was known to be crucial for its anti-cancer activity. However, the aggressiveness (nucleophilicity) of the nitrogen atom determined the reactivity of the drug, because the nitrogen needed to flip around and cause the chloride leaving group to dissociate. If the nitrogen was too weak to do this, the molecule would remain in an inert, inactive state.[26]

Brock and his colleagues created a series of compounds that had a weak nitrogen, unable to flip into the active drug molecule. They hoped that these molecules could serve as pro-drugs, entering tumor cells where they would be activated. One of these, cyclophosphamide (saiy-klo-FOS-fah-mayd), was found to reduce the size of tumors in rats and mice.

One of the critical characteristics of any drug is its therapeutic window, which is the amount of drug needed for a beneficial effect compared to the amount of drug that produces intolerable side effects. If a drug has a beneficial effect at one dose level but doubling the amount of the drug is fatal, we would say this drug has a narrow therapeutic window, meaning that there is a small range of doses that gives a useful effect. Since each patient will respond a little differently to the drug, this narrow range makes it likely that some patients will experience undesirable side effects, perhaps even fatal ones. On the other hand, a drug that doesn't cause side effects at any dose would have an enormous therapeutic window and would be much preferable. Cyclophosphamide was an improvement over other reactive drugs, such as the mustard agents, because cyclophosphamide was found to have a greater therapeutic window.[27]

Ultimately, in a wide range of clinical studies, cyclophosphamide was found to have anti-tumor activity in many cancers, including chromic lymphocytic leukemia, granulocytic leukemia, numerous lymphomas and sarcomas, multiple myeloma, and breast, lung, and cervical cancers.[28] It is now marketed under the brand names Procytox, Cytoxan, and Neosar, among others, and is a mainstay in standard cancer chemotherapy regimens for many tumors.[29]

Treatment with cyclophosphamide was a transformative event in many patients' lives, and it was adopted internationally. One patient thus affected was an otherwise healthy 76-year-old Japanese woman who was admitted to the Kumamoto University Hospital in 1994 because of an elevated white blood cell count that had persisted for the preceding 11 months.[30] Upon examination, she was found to have an enlarged liver and spleen. The elevated white blood cell count was subsequently attributed to mantle cell leukemia (MCL). She was treated with cyclophosphamide daily for a month. Her white blood cell count decreased and the enlargement of her spleen disappeared. Six months later, her white blood cell count remained normal, as did her spleen size. She was in remission. Her remission continued for over four years without any relapse or symptoms. Her prognosis remained excellent at the end of the study. Her life was transformed by a small molecule named cyclophosphamide.

FROM VITAMIN DEFICIENCY TO A CANCER DRUG

Cyclophosphamide is a type of cancer drug that generates a reactive molecule that can damage DNA and proteins; therefore, there is no specific protein target for this drug. However, other early cancer drugs were more specific in their mechanism of action. One such drug was developed by Sidney Farber.

Farber, a pediatric pathologist, was the leader of a group at Boston's Children's Hospital that was exploring treatment ideas for children with leukemias, or white blood cell cancers. Typical of such leukemia patients was one we will call "Benjamin" (not his real name) a 5-year-old child diagnosed at the University of Michigan. Benjamin had a high white blood cell count, and many tumor cells present in his blood; he had a poor prognosis and likely would not live long—a devastating piece of news for anyone and incomprehensible for a 5-year-old. In that era—the first half of the twentieth century—nearly 100% of pediatric leukemia patients died of the disease. Benjamin's life would be dramatically altered by the perseverance and research of Farber and his colleagues.

It was known that deficiency in the vitamin folic acid could result in bone marrow suppression, which is a condition in which the production of blood cells is impaired. In normal individuals such bone marrow suppression is problematic, causing anemia and difficulty in responding to infections. However, in patients with leukemia, bone marrow suppression could be helpful, as it might limit the growth of the leukemic blood cells.[31] In addition, Farber had observed that folic acid could increase the growth of leukemia cells, a fact confirmed by Robert Heinle and Arnold Welch in 1948.[32]

Farber assembled a clinical team to investigate whether folic acid antagonists, compounds that prevented folate synthesis inside cells, would be effective at treating leukemias. In the spring of 1948 this team treated 16 children with acute leukemia, using aminopterin, an anti-folate they obtained by working with a group of chemists at Lederle Laboratories, which became part of Wyeth, which was in turn acquired in 2009 by Pfizer. Farber and his team found that 10 of the 16 children showed striking clinical improvements after treatment with aminopterin. Shortly thereafter, they began testing a second antifolate named methopterin, which had even stronger anti-cancer activity and fewer side effects. The name of this compound was changed to methotrexate (meth-oh-TREX-ayt) and it went on to become one of the most widely used anti-leukemia drugs.[33] It also has been used in treating autoimmune diseases, because it can effectively block the growth of activated immune cells. Such diseases include Crohn's disease, psoriasis, rheumatoid arthritis, and inflammatory diseases of the intestines, skin, and joints, respectively.

Unexpectedly, the slowly emerging revolution in treatment strategies for pediatric oncology patients created tension between those who believed cancer would ultimately be cured with drugs, and those who thought treating cancer with drugs was a naïve and dangerous dream. Farber's team reported that some of the other staff in the hospital were actively hostile to their efforts to treat children with drugs. Most of the children were expected to die, so most of the staff became accustomed to the approach of simply making the children comfortable and watching them pass in peace. However, to carry out the drug treatment regimen, oncologists needed to extract bone marrow repeatedly, which was a painful procedure. Also, the drug treatment carried serious toxicities, which was unacceptable to those physicians who were certain of the treatment's likely failure.[34]

One of Farber's clinical fellows, Robert Mercer, reported being called into the office of Charles Janeway, chief of the medical service and in charge of these other physicians, and grilled on his clinical approach to pediatric oncology patients. Mercer explained the strategy behind the drug treatment approach, the substantial progress already being made, and his hope that anti-folates could change the paradigm for treating pediatric oncology patients. Janeway was ultimately persuaded, and in time the hostility of the other physicians subsided. These battles were but one emotional challenge these pioneering oncologists faced. Mercer, for example, recalled his profound sadness at seeing many terminally sick children die, despite his best efforts.[35]

Reports of successfully treating pediatric leukemia patients generated a surging interest in oncology, and Boston's Children's Hospital became a center for drug therapy of childhood leukemias. In 1947 Farber established a foundation to focus efforts on the treatment of childhood cancers, which he named the Children's Cancer Research Foundation. This foundation began using the name "The Jimmy Fund" in honor of a young patient who went by the name Jimmy and was the focus of a radio broadcast. The patient's actual name was Einar Gustafson, and he was a patient of Farber's when he was 12 years old. In 1983 the name of the research foundation was changed to the Dana Farber Cancer Institute; it is now one of the leading cancer centers in the world.

One of the three clinical Fellows on Farber's team was James Wolff, who subsequently left Boston for New York City, where he joined the staff at Babies Hospital, now known as Morgan Stanley Children's Hospital of NewYork-Presbyterian. He built a group with great expertise in pediatric oncology. There is now a named professorship in the Department of Pediatrics in Wolff's honor.[36]

In 1998 that 5-year-old patient, Benjamin, diagnosed at the University of Michigan in 1952 and facing near certain death, was rediscovered. Yaddanapudi Ravindranath, a clinician, was following up on long-term survivors of childhood leukemia. He had not had great success, but the name of that one child had remained with him.

One afternoon he found himself at a golf fundraising event for a children's hospital, and he heard the name. He approached the adult Benjamin and confirmed that the 5-year-old boy from his records had grown into the man he had found. Benjamin, facing near certain death, was treated with an anti-folate drug and survived for over 45 years.[37]

SYSTEMATIC DISCOVERY OF CANCER DRUGS

The compelling results with anti-folates and cyclophosphamide emboldened researchers to think about systematically searching for drugs that could be used to treat cancers. At this time, there wasn't a significant amount of anti-cancer drug discovery occurring in the nascent pharmaceutical industry, aside from the few examples mentioned, because it was widely considered a lost cause. Most of these early efforts took place in academic and government settings, and a few in collaboration with the limited number of pioneers in industry. In addition, there was little knowledge of which proteins would make suitable targets for cancer drugs. Instead, the focus was on finding drugs that could kill cancer cells, without regard to the proteins with which these small molecules might interact.

With a newfound enthusiasm for systematic cancer drug discovery, a number of researchers began collecting soil samples for testing against cancer cells. In this approach the microorganisms present in each soil sample were grown in a culture, or liquid broth, in the laboratory, and then the small molecules present in the culture broth would be extracted using an organic solvent. Each such extract would then be tested for its ability to prevent the growth of tumor cells in a plastic dish in the laboratory. Before long, researchers also began creating these extracts from other natural sources, such as fungi, plants, and marine sponges.[38]

Crucial to all of these screening efforts was the ability to grow tumor cells in the laboratory under defined and reproducible conditions. This technology, known as cell culture, had begun in the late nineteenth century with the work of Sydney Ringer, a medical doctor and Professor of Medicine at University College in London. In 1882 Ringer published a study in the *Journal of Physiology* in which he studied the effects of different solutions and blood constituents on the beating of an isolated heart.[39] Then, in 1885, Wilhelm Roux was able to remove part of an embryonic chicken and maintain it in culture for several days.[40] These early researchers were intrigued by their ability to grow parts of animals outside the body. In the period from 1907 to 1910, the embryologist Ross Harrison at Johns Hopkins Medical School found that he could keep fragments of embryos alive in this way.

It was Alexis Carrel, a surgeon at Rockefeller University, who created the first cell culture, as opposed to an organ culture, where he was able to grow human and mouse cells in a dish outside of an animal. This was a shocking pronouncement, and one that was greeted with skepticism and controversy. This discovery had the kind of psychological effect on physiologists that Wohler's synthesis of the organic compound urea from inorganic materials had had upon chemists—it seemed totally implausible. Nonetheless, over time this methodology was adapted to a wide array of different cell types and culture conditions, allowing for the detailed study of cells in a controlled setting outside of living organisms. In time cell culture would become an essential part of the drug discovery enterprise.[41]

Monroe Wall and Mahsukh Wani used cell culture methods to screen plant and soil extracts for anti-cancer activity. Wall directed a large screening program within the U.S. Department of Agriculture in Philadelphia, in which he tested thousands of plant extracts that had been obtained by plant biologists, cataloged, and sent to him. One such plant that had been collected was the *Camptotheca acuminata*. This deciduous tree, which is found in Southeastern China, grows to about 60 feet.

Of 1,000 plant extracts tested, only the extracts of the *Camptotheca* tree had anti-tumor activity in the assay system used. To identify the active compound in this extract, Wall obtained 44 pounds (20 kg) of wood and bark from the *Camptotheca* tree and separated it into a series of different fractions using a variety of solvents and separation technologies, finally obtaining a tiny bit of the active molecule, which he named camptothecin. Camptothecin itself was too toxic for human use, but eventually two analogs were synthesized, irinotecan, and topotecan, that showed powerful anti-tumor properties in humans with acceptable toxicity. Irinotecan is now used to treat patients with colon cancer, while topotecan is used for patients with ovarian and small cell lung cancers, among others.[42]

Similar efforts over the next 30 years led to the discovery of etoposide, doxorubin, and paclitaxel, all of which are now widely used chemotherapeutic drugs for treating a variety of cancers.[43] These drugs were discovered by systematically testing natural extracts for their effects on tumor cells growing in a laboratory. The discovery of paclitaxel (pak-lih-TAX-el) was particularly dramatic and began when the United States Department of Agriculture collected a sample of bark from the Pacific yew tree in 1962. An extract of this bark was found to prevent tumor cells from growing in plastic dishes in the laboratory, suggesting that the bark contained a molecule of interest. The active molecule was successfully purified and its chemical structure determined in 1971, revealing an architecture of unprecedented complexity.

Susan Horwitz, working at the Albert Einstein College of Medicine in New York City, discovered that this new molecule had a new mechanism of action—it attached itself to microtubules, the railroad tracks of cells that allow proteins to move from one part of the cell to another, and prevented these microtubules from being reorganized, which is needed for their normal function. In 1983 human clinical trials of paclitaxel were initiated. The drug showed a marked ability to slow ovarian cancer progression and caused complete remissions in some patients. Initially, it was not possible to provide the needed amount of drug by extracting it from the bark of the Pacific yew tree, as this would have shortly led to the destruction of most of these trees. However, in time, sustainable routes to making the drug were developed; in 1992 it was approved for use in ovarian cancer, and it was later approved for advanced breast cancer, among other tumor types. It took 30 years from the initial collection of yew tree bark until this drug was approved for use in cancer patients, highlighting the long time horizon for developing many drugs. Paclitaxel, also known as Taxol, has now become a critical addition to the anti-cancer drug arsenal.[44]

SERENDIPITOUS DISCOVERY OF A CANCER DRUG

One important cancer drug was discovered in an unusual manner.[45] This case involved Barnett Rosenberg, who was a physicist working at Michigan State University in 1965. He was interested in testing the idea that magnetic fields could affect cell division in bacteria. To test this notion, Rosenberg inserted platinum electrodes into a solution of growing *Escherichia coli* bacteria. He noticed an unusual change in the morphology, or shape, of the bacteria, which he eventually realized was not due to a magnetic field but was instead caused by the formation of a platinum-containing molecule at the site of the electrode.

Rosenberg had resynthesized Peyrone's chloride, which was first made in 1845, and which has a central platinum atom surrounded by two chloride ions and two amino (NH_3) groups. Shortly thereafter, this compound was found to cause tumor regression in mice. By 1971 it was renamed cisplatin and tested in patients, and in 1978 it was approved for the treatment of cancer patients by the U.S. Food and Drug Administration (FDA). More effective platinum compounds, such as carboplatin and oxaliplatin, have been invented over the years and now represent an important group of anti-cancer drugs.[46] A dramatic anecdotal demonstration of the power of platinum-based drugs was provided in the treatment of Lance Armstrong, the professional cyclist who was diagnosed with testicular cancer in 1996. His cancer had spread dangerously to his brain and lungs. Armstrong was treated with cisplatin, as well as other chemotherapy drugs, and was ultimately cured of his cancer. He subsequently succeeded in winning the Tour de France, an extremely competitive bicycle race, seven times, from 1999 to 2005.

Most currently used cancer chemotherapy drugs can be divided into five major groups. First are the platinum-based drugs, such as cisplatin, which act by damaging DNA. Second are a variety of natural-product-derived compounds, such as topotecan, etoposide, doxorubicin, and other related molecules, which also act by damaging DNA. Third are alkylating agents, such as mustards and cyclophosphamide, which again act by damaging DNA. Thus three of the major classes of cancer drugs act primarily by causing damage to cellular DNA.

Fourth are the anti-metabolites, such as methotrexate, which largely function to prevent the synthesis of the building blocks of DNA; therefore, in a sense, the ultimate target of these drugs is also DNA. Fifth are the anti-microtubule agents, such as Taxol, which disrupt microtubule functions crucial for allowing DNA replication. Again, the ultimate effect of these compounds is to prevent proper DNA replication and therefore cell division. It is remarkable that in many ways these drugs all target the same fundamental molecule in the cell—DNA—yet they have quite distinct efficacies in specific cancers.[47]

Before the discovery of these chemotherapy drugs, there was little hope of being able to treat cancer. Step by step, the discovery of each of these classes of cancer drugs changed the notion of cancer from an unbeatable malady to one that can in time be cured. Cancer drug discovery is now a focus of almost every major pharmaceutical company and many biotechnology companies, as well as numerous groups in government and academia. This golden age of chemotherapy succeeded in fundamentally altering the perspective on cancer, so that it is now seen as a disease that is potentially curable.

FROM HORMONES TO CANCER DRUGS

Although chemotherapy drugs changed the face of cancer treatment, by the late 1980s researchers had largely reached the limits of these drugs. They act by targeting the rapid rate of cell division characteristic of tumor cells. But normal cells that proliferate rapidly are also injured by these drugs. This damage causes the commonly known and devastating side effects of cancer chemotherapy: bone marrow division is suppressed, causing a weakened immune system and anemia; the cells lining the gastrointestinal tract are killed, causing nausea and mucositis, or damage to the mucous membranes of the digestive tract; and hair follicle cells are destroyed, causing hair loss. Moreover, some tumor cells grow slowly and insidiously and are therefore not targeted by these agents. It became apparent that a better approach would be required to make further improvements in cancer treatment.

One strategy that gradually developed for treating cancer without targeting DNA was based on preventing the hormone-driven growth of some tumors. This thread of research began in 1896, when George Beatson discovered that removing the ovaries from some women with breast cancer could prevent further growth of their breast tumors. However, only 1 in 3 patients responded to this invasive procedure. Elwood Jensen showed that a radioactively labeled estrogen (estradiol) could bind to the breast tumors that responded to removal of the ovaries. In other words, the responsive breast tumors harbored a protein that could bind to estradiol. This protein became known as the estrogen receptor. Breast tumors with this protein present were driven to grow larger by estrogen secreted from the ovaries; hence, removing the ovaries prevents the further growth of these tumors. Breast tumors without the estrogen receptor are driven to grow by other mechanisms and don't respond to removal of the ovaries.[48]

In an independent line of investigation, Leonard Lerner at Merrell Laboratories began exploring the activity of triphenyl (TRI-fen-ul) compounds in the 1950s. He found that these compounds blocked the activity of estrogen in all species tested, which suggested that they could be used as a morning-after contraceptive. To his dismay, these compounds had the opposite activity in humans—namely, they stimulated ovulation, preventing their development as contraceptives.

Subsequently, Arthur Walpole at ICI Pharmaceuticals, which is now part of AstraZeneca, discovered a related molecule, ICI46,474, which was also abandoned as a contraceptive. However, the work of several researchers, including Lars Terenius, Arthur Walpole, and Craig Jordan suggested that anti-estrogens such as ICI46,474 might be useful in the treatment of breast cancers that contain the estrogen receptor. This compound, eventually renamed tamoxifen, and a related compound called raloxifene, as well as other analogs, went on to became major additions to the anti-cancer drug panel. They are widely used both for the treatment of estrogen-receptor-positive breast cancers, as well as for the prevention of breast cancer in women at high risk of developing the disease.[49]

Antagonizing the binding of estrogen hormones to the estrogen receptor is not the only way to block the tumor-promoting effects of these hormones. In 1973 Angela Hartley Brodie published her initial studies attempting to create small molecule inhibitors of aromatase, an enzyme essential for the synthesis of estrogen hormones in the adrenal gland of post-menopausal women. These studies built on decades of work by many investigators elucidating the biological functions of steroid hormones and the precise biosynthetic pathways used to construct them. There is now an important class of aromatase-inhibiting drugs that are used for treating breast and ovarian cancers that depend on estrogens for their growth. These drugs include exemestane, anastrozole, and letrozole, among others.[50]

This concept of targeting hormone-dependent tumors is not limited to breast cancers. Some cancers of the prostate in men are dependent on testosterone for their growth, synthesized by the testicles. Drugs that block the synthesis of testosterone, called luteinizing-hormone-releasing-hormone (or LHRH) agonists, have been developed as an alternative to castration, which is safe and effective but obviously not a psychologically attractive option for men. These LHRH agonists are the equivalent of the aromatase inhibitors for breast cancer. There are also anti-androgens that block the effect of testosterone at its receptor, which are conceptually similar to tamoxifen for breast cancer. These therapies have had a major impact on the treatment of prostate cancer, particularly when the disease is not locally confined and is less amenable to local treatment strategies such as radiation or surgery.[51]

Hormone-related therapies have also been developed for patients with other types of cancer. For example, some tumors involved with the hormonal interface between the endocrine system and the nervous system (called neuroendocrine tumors) can be successfully treated with somatostatin analogs, which are peptide hormones that regulate these tumor cells. Several such drugs have been approved for the treatment of these tumors.[52]

However, this strategy of targeting hormone-dependent cancers has likely reached its limits. Most cancers are not hormone-dependent in their growth; thus many tumors will not be subject to treatment with this approach. Moreover, targeting hormonal signaling carries the risk of significant side effects caused by disrupting the normal functions of these hormones.

ANGIOGENESIS AND BEYOND

Cancer drug discovery began by targeting one observable property that distinguishes tumor cells from normal cells—their rapid rate of proliferation—and then productively mined this difference between normal cells and tumor cells until the limits of toxicity were reached. Then a second distinction between some tumor cells and normal cells—hormone-dependent growth—was targeted. Again, this property of tumor cells was productively mined to create new, effective drugs. Yet, again, this approach ultimately reached its limit. Are there other gross differences between tumor cells and normal cells that can be exploited for therapeutic purposes? There is at least one other property that has been studied—angiogenesis.

Judah Folkman pioneered the notion that tumor cells require the ability to stimulate new blood vessel growth in order for the tumor to grow to a macroscopic size; this process is known as *angiogenesis*. Folkman argued that blocking angiogenesis could be a way to starve tumors and prevent their growth. As early as 1972, Folkman showed that tumors could not grow if they were deprived of new blood vessels.[53] Over a 40-year period, Folkman and the members of his laboratory explored this concept in detail and discovered a number of inhibitors of tumor angiogenesis, including interferon, TNP-470, angiostatin, endostatin, and thalidomide.[54] Some of these agents had a modest clinical benefit in specific tumors, but the field of angiogenesis inhibition was finally validated when Genentech's drug targeting vascular endothelial growth factor (VEGF) was approved in 2004. This drug, called bevacizumab, or Avastin, is a type of large protein known as a monoclonal antibody that blocks the angiogenesis activity of the secreted protein VEGF. Avastin was approved by the FDA for the treatment of metastatic colon cancer, non-small-cell lung cancer, kidney cancer, and breast cancer. However, in the summer of 2010, an advisory panel to the FDA, the government agency responsible for approving drugs for use in patients, recommended that the FDA revoke the approval of Avastin for breast cancer. This recommendation was made after two large clinical studies showed no substantial benefit in breast cancer patients.

As you might imagine, there can be side effects caused by blocking normal blood vessel formation, such as abdominal pain, constipation, and other gastrointestinal problems. Moreover, opinions on the potential breadth of anti-angiogenic therapy vary. Although these drugs are effective in treating colon cancer, some researchers believe that they will not be dramatically effective for other types of cancer.[55] Thus, while angiogenesis inhibitors certainly have a role to play in the cancer drug armamentarium, we will need to search elsewhere for dramatic improvements in cancer treatment.

Each drug that has been discovered to date represents in many ways a lasting memorial to a scientist who doggedly pursued development of the drug in the face of the near universal failure in the drug discovery enterprise. Each of these scientists, and their associated teams of researchers, persevered and brought a new drug to patients. But there is a limit to what one drug can do on its own, just as there is a limit to what one person can do on their own. What if we could combine these drugs in just the right ways, creating powerful drug combinations? In the next chapter we will explore the possibility that by working together, perhaps these drug cocktails might accomplish what no individual drug can do on its own.

4

A NEW COMPANY CREATING DRUG COMBINATIONS

One potential solution to the drug discovery shortage is to find new uses of existing drugs. Since it is extremely difficult and expensive to find new drugs, perhaps the research community should make the most out of the precious few that have successfully run the gauntlet from the lab to the pharmacy. In other words, if the pessimists are right in saying that only 2% of proteins are druggable, we should maximize the usefulness of the drugs that have been found to target one of these rare druggable proteins.

There are two strategies for making better use of existing drugs. First, we can try to find new diseases to deploy existing drugs against—sometimes it turns out that a drug can be used to treat a disease that it was never designed for. A classic example of this is the discovery of Pfizer's drug Viagra. Viagra is a small molecule that inhibits the enzyme PDE5 (phosphodiesterase 5, FOS-foh-dai-ESS-ter-ays). Viagra was originally designed to treat high blood pressure, but during clinical testing in patients, it was unexpectedly found to induce erections in men for an extended period of time. The drug was successfully developed and marketed for the treatment of erectile dysfunction, creating a cultural icon.[1]

A second strategy for making better use of existing drugs is to use combinations. In some cases a combination of two drugs can provide a therapeutic benefit when either one by itself would be ineffectual. For example, most cancer therapies given to patients consist of combinations of two, three, or four different drugs, because extensive clinical testing has found these combinations to be more effective than the individual drugs.[2]

Some drugs are already marketed as combinations per se. One of the most effective antibiotic drugs, augmentin, is a combination of two small molecules, amoxicillin, and potassium clavulanate. Amoxicillin is used to treat bacterial infections. Some bacteria express an enzyme called beta-lactamase that destroys amoxicillin and allows these bacteria to become resistant to amoxicillin. Potassium clavulanate inhibits beta lactamase and thereby blocks this resistance mechanism. By combining these two drugs, it is possible to create an antibiotic regimen that is effective against a broad range of bacteria, and it is less likely that bacteria will develop resistance to the combination. By using this drug combination, it is possible to increase the overall therapeutic benefit to patients.[3]

THE BIRTH OF A NEW COMPANY

In 1999, when I had just finished graduate school, I founded a biopharmaceutical company that sought to identify combinations of existing drugs that could be effective in diseases that the original drugs were never used for. I had always wanted to found a company, ever since I was quite young. I can't recall the first time I had this ambition, but I trace it back to my mother's repeated efforts at making me think and function independently, encouraging me to set and realize my own goals.

In high school my friend Rob and I contemplated starting a company that would sell medical devices, because we imagined there would always be a demand for such products. In fact, we had a two-stage plan to start this company because we realized we didn't have the funds in hand to start a medical device company at the age of 15. First, we would start a clothing company that made and sold T-shirts;

we had experience doing this and knew that the barrier to entry was low, so we could get started right away. Next, we would reinvest our profits from the T-shirt company into a medical device company. We never quite got to the point of launching this start-up, but we enjoyed planning it.

In college I took my interest to a more sophisticated level by taking a course on starting a new venture at Cornell's Business School; in this context, I wrote a business plan for starting a company that would lease scientific equipment to research labs. I strongly considered pursuing this venture when I finished my undergraduate studies, but ultimately I decided to pursue a PhD and postpone this ambition.

While I was immersed in my graduate studies, I spoke frequently about my interest in starting a company with my friend and fellow graduate student Mike Foley, who shared this entrepreneurial passion. We would sit in the laboratory conducting experiments and chat about ideas for new biotechnology companies. Some of the best scientific ideas are hatched by students and postdocs in between the actual experiments they are running. We often found ourselves with 5 or 10 minutes in between steps of a protocol. This wasn't enough time to go read a journal article or to go for coffee, so we found ourselves standing at the lab bench either listening to the radio (usually National Public Radio) or talking to whoever wandered by. In these random encounters, surprisingly effective insights, plans, and collaborations can be hatched. Architects who design laboratory buildings now try to design in these serendipitous encounters. In such moments Mike and I hatched our plans to start a new company.

As soon as we received our doctoral degrees, we began moving ahead in earnest with several ideas for a start-up company. The whole episode is remarkable to me in retrospect—we were two young guys with no business experience and no breakthrough patents or ideas, just an overwhelming passion to start a successful biotechnology company. Fortunately for us, this was 1999, when 19-year-old students were starting multimillion-dollar Internet companies. Compared to some Internet start-ups, I suppose we were a relatively conservative investment. On the other hand, biotechnology start-ups need a lot more money than Internet start-ups, because lab equipment and scientists are expensive, making it more challenging to create such a venture.

At this time, in my graduate school research I had become focused on the idea of assembling a collection of all the drugs that had ever been approved for use in humans. My doctoral advisor, Stuart Schreiber, had planted the seed for this idea in my mind. He had been thinking about the future of research at the interface of chemistry and biology, and he articulated a vision for ultimately creating a small molecule that could bind to each human protein.[4] This was a compelling goal to aspire to—a complete collection of small molecule modulators of human protein function. As I began thinking about this vision I realized that instead of waiting years for all of these precious small molecules to be discovered, we could start right away by assembling a collection of the 20,000 drug products that had been approved for use in humans by the U.S. Food and Drug Administration (FDA). However, as I investigated this notion I realized that these 20,000 different products actually represented only 1,357 different small molecule drugs sold under different names.[5] Moreover, I began to realize that these 1,357 drugs collectively target only a few hundred proteins.[6]

Nonetheless, even with its small size I thought this would be a powerful collection. These approved drugs have a rich body of knowledge associated with them, making them easier to work with and easier to develop for new uses. We already know how to manufacture and formulate these drugs, how to give them to people and animals so they will have a medical benefit, and we know a lot about the mechanisms by which these drugs affect proteins and cells. The traditional drug discovery approach of selecting a protein first and then seeking a completely new compound that binds to this protein and that can be optimized to work in patients is slow, expensive, and failure-prone. I realized that the approach of working with existing drugs greatly increased the chance that the drugs would ultimately be delivered effectively to animals and patients.

I had also become interested in the idea of combining existing drugs in new ways. During my PhD thesis I began testing pairs of compounds in combination to see what effects they would have compared to the individual compounds.[7] I had seen some surprising effects in which the combination of two compounds was either dramatically more effective or less effective than the individual molecules. This intrigued me as I began to wonder how common these synergistic and antagonistic effects were and if they could be harnessed for drug discovery.

Mike Foley was a perfect partner for a biotech start-up venture. Mike had worked for a decade in the pharmaceutical industry, first at Bristol-Myers Squibb and then at Glaxo Wellcome (now GlaxoSmithKline), prior to starting his doctoral work in Stuart Schreiber's lab. Glaxo held a competition for a fellowship, in which they would pay for the cost of one of their employees to obtain a PhD in a top graduate program. Mike applied and won this fellowship competition, and before long, he found himself and his family transplanted from Chapel Hill, North Carolina to Cambridge, Massachusetts. In time Mike would finish his PhD and go on to become one of the first independent research Fellows at a new Harvard institute at the interface of chemistry and biology, founded by Stuart Schreiber and Tim Mitchison, named the Harvard Institute of Chemistry and Cell Biology (ICCB). Mike would then become cofounder and Vice President of Chemistry at Infinity Pharmaceuticals and would cofound Forma Therapeutics. But before all of that, there was just the two of us, dreaming about starting our first biotechnology company.

Mike is rigorous and analytical, especially in terms of designing research plans. He could design the optimal process for every aspect of a new company's existence. From his perspective, starting a company was the next logical step in his transformation from a company man to an independent researcher. At Glaxo he had a secure, easy job with a relaxed lifestyle, but he hungered for greater control of his destiny and the ability to change the world, at least in some small way. Together we hoped to do something completely different from what others in the pharmaceutical industry were doing.

As Mike and I were hatching our plans, our friend and colleague Curtis Keith would listen in and offer suggestions. I don't think Curtis thought we were serious at first, but he enjoyed hearing about our plans. Curtis is a sharp thinker, who can quickly analyze any piece of scientific data and design the next crucial experiment. He can rapidly cut through to the core of any scientific issue. As Mike and I refined our ideas, Curtis had the foresight to introduce us to Alexis Borisy, a former student from the same lab. All four of us, Mike, Alexis, Curtis and I, had done our graduate training in Stuart Schreiber's laboratory in the chemistry department at Harvard. Alexis had spent a few months in the lab but ultimately decided to pursue consulting, especially related to pharmaceutical strategies. He became skilled at analyzing the strategic directions of pharmaceutical and biotechnology companies and made contacts in the investment community. Alexis independently had the idea of starting a venture.

I remember the day in October 1999 when Mike, Alexis, and I finally got together in the basement of a bohemian café called Curious Liquids, on Beacon Street across from the State House in Beacon Hill. We sat in large, overstuffed, dusty chairs, sipping frothy coffee beverages, and talked about starting a company.

Alexis had also been thinking about combinations of molecules and wondered about starting a company on this basis. I persuaded him that random screening of combinations was unrealistic because of the vast number of possible combinations and the huge challenge of turning two new compounds into marketed drugs. However, by using the collection of existing drugs I had already begun assembling, we thought we could find new combinations of drugs with unanticipated and powerful effects on disease. The notion was to use the kind of automated robots I had been working with in my graduate research to rapidly test millions of pairwise combinations of approved drugs for their effects in cells. If we could find unexpected synergies of drugs for use against diseases that the individual drugs had never been developed for, we might be able to develop and market these new combinations effectively. Before long, we convinced Curtis to join us in our new undertaking.

I was fortunate that my wife, Melissa, was a medical student and was willing to support our new venture with free consulting. She pored through pharmacology and drug handbooks to find many examples of effective drug combinations as well as toxic drug combinations. We wanted to see whether it was necessarily the case that synergistic drug combinations would also have synergistic toxicity, which was an initial objection to our plan that was raised by several people. Through Melissa's searches, we came to the conclusion that there were many different effective combinations in clinical use and that combination toxicity was a relatively rare event and independent of any beneficial synergy. This suggested that we should find many useful combinations that didn't have unexpected toxicity. We decided to move ahead with this concept for a new company.

Fortunately, we were naïve at that point about what it would take to start a biotechnology company. We forged ahead, not realizing the challenges to come. The first problem, of course, was funding. Alexis had made a number of contacts with so-called high net worth individuals—people who have enough money to invest their own funds in start-up companies. We approached one of these potential angel investors, Jacob Goldfield. We were able to convince Jacob to invest in our idea, based on the possibility that this could be a powerful and inexpensive new route to drug discovery. I doubt we would have made much progress in starting this company without Jacob's willingness to jump in at such an early stage. To be an angel investor in biotechnology (the name for this sort of early stage investment), you need to have a long time horizon and a tolerance for risk. Building a biotechnology company, as we eventually discovered, takes hundreds of millions of dollars and nearly a decade.

We used the initial investment of $2.5 million to file a series of patent applications on our idea of screening for new combinations of existing drugs and to set up a lab to try the big experiment—combining existing drugs and looking for unexpected synergies. I thought it would be difficult to find people to work in a brand new start-up company with no history, no established procedures and policies, and no guarantee of any future. Surprisingly, finding employees was not that hard, even in the year 2000, when the economy was sizzling. It was much harder to find lab space for the company than it was to find employees. For a while we couldn't even get a call back from the laboratory real estate agents we tried contacting, and eventually we were told there was absolutely no lab space in all of the Boston/Cambridge area. Mike and I then went door to door, literally knocking on biotechnology company doors and asking if they had a few spare lab benches we could rent.

Eventually, we found a small, dilapidated basement lab situated in South Boston. Within a short period of time, we hired a dozen researchers and purchased all the lab equipment we would need to do this research: specialized incubators and workstations for growing and working with human cells, robotic instruments for mixing together thousands of different combinations of molecules, detectors for quickly measuring the effects of these combinations on cells, and most crucial of all, a large number of approved drugs. We had a gleaming new lab ready for operation. Unfortunately, it was located in a dingy basement next to Liquor Land in a section of South Boston frequented mostly by those who are down on their luck. But we had started our new company.

We had a difficult time coming up with a name for our company, probably because we were so inexperienced. We wanted the name to be unique and meaningful as well as memorable—something to last for the ages. We spent many hours discussing possible names. We considered naming the company after ourselves but decided that if the company ultimately were to be sold, it would be uncomfortable to have our names floating around on a venture we were no longer associated with; in addition, it sounded too much like a law firm. At one point we were enthusiastic about the idea of Einsteinium-252. Einsteinium is element number 99, and its most stable isotope has a mass of 252. We liked the idea of a company based on the creativity of Einstein, and one with a little bit of zaniness associated with it. We imagined a stylized Einstein hairdo as the company's logo. Ultimately, though, others didn't like the name, as they thought it sounded like a physics or nuclear weapons company instead of a biotechnology company.

We also considered more prosaic names, such as Combination Sciences, but these lacked personality. Finally, we settled on the name CombinatoRx, which was a subtle play on the branch of mathematics that deals with combinations, which is called combinatorics. By changing the ending of this word combinatorics to Rx, which is medical shorthand for the word *prescription*, we created a hybrid word that sounded like a biotech company, but was pronounced like "combinatorics."

We thought it was clever, but nobody else ever liked the name, nor could they pronounce it or spell it. In the future, if you are starting a company, I recommend that the name be easy to pronounce and easy to spell; investors will like it better—they don't have time for wordplay. Even when the company was fairly well advanced, employing at least 50 people and having raised over $18 million, we continued having discussions with the board of directors over whether we should change the name. We almost paid a consultant $100,000 to come up with a better name, but in the end we finally all agreed to leave the name as it was and live with it.

FINDING NEW DRUG COMBINATIONS

Indeed, we did find striking synergies. One of the earliest combinations we found, and one that was subsequently tested in human cancer patients, was the combination of the antipsychotic drug chlorpromazine with the anti-parasitic drug pentamidine.[8] Alone, each of these drugs had a modest ability to inhibit tumor cell growth. Together, however, they had a substantial ability to block the growth of tumor cells in a dish. This combination was also effective in mouse cancer models. In these tests we saw the same theme—the individual drugs had modest anti-tumor activities, but the combination was consistently effective at blocking tumor formation in mice.

We moved this combination through the drug development process and eventually obtained approval from the FDA to test it in patients. Unlike most cancer therapies, it didn't cause the hair loss, nausea, or the other side effects that are usually caused by chemotherapy. However, we did see the frustrating side effect of somnolence—patients would drop off to sleep shortly after being given an infusion of the drug. This made it inconvenient, because they would have to be given a place to sleep after receiving the drug, and this ultimately reduced enthusiasm for the drug. But the bigger issue was that when we tested this combination drug in patients, we didn't see the striking anti-tumor effects that we had seen in cells and in mice. This is one of the major problems in drug discovery. Cells in a plastic dish and mice in the lab are just not the same as patients in the clinic. Despite all of the advances in basic research, we still have a long way to go to effectively model cancer in animals, and even more so in cell culture.

Nonetheless, as these studies were proceeding, we sought additional financing to expand the company and seek combinations that might be effective against other diseases. We met with a large number of venture capitalists (VCs, as they are known). These investors see presentation after presentation from a long line of entrepreneurs, each thinking they have the next great idea. The VCs are responsible for hundreds of millions of dollars that they have raised from other investors; they need to decide how to invest this money to earn the best possible return for their investors.

From the perspective of those of us seeking financing, we go from VC office to VC office in a seeming endless parade, telling the same story over and over and answering the same questions. It can be a frustrating experience on both sides. On the other hand, many venture investors are sharp thinkers and former scientists, and it can be exhilarating to discuss your scientific ideas with such sharp minds. We found that our plans were solidified and refined through these discussions.

This raises an interesting point about scientific ideas. There is a popular notion about new ideas in science springing forth from a great mind fully formed in a dazzling eureka moment. In my experience this is not accurate. There are certainly sudden insights and ideas that appear to you from time to time. Many times, of course, a little further thought makes you realize it is really an absolutely terrible idea after all and could never work. But even when you have an exciting new idea, it begins as a raw, unprocessed idea. Some digging around in the literature will allow you to see what has been done before, and whether this idea is novel and likely to work. If the idea survives that stage, it is still full of problems and flaws, in both the content and the style of presenting it. However, the real processing comes from discussing the idea, informally at first, with students and colleagues. Then, as it is presented in seminars, each audience gives a series of comments, suggestions, and questions that help mold the idea into a better, sharper, and more robust proposal. Finally, there is the ultimate process of submission for publication, review and revision, and finally acceptance, where there is a final culling of any weaknesses. The scientific process is a social process, where you refine your ideas through repeated discussions and presentations. At the end of this series of tests, the idea that remains compelling and viable is likely to stand the test of time. Funding a new company is similar—the proposal gets refined after each presentation, and new objections are the basis of alterations and revisions to the plan. In this way, searching for financing actually helps build a strong idea and a strong company.

Ultimately, though, the goal of such discussions is to obtain actual funds that can launch your idea. From this perspective, searching for financing is a lot like dating, and even a lot like high-throughput screening of chemicals. You test a lot of candidates, and you try to go as fast as you can and find a few winners that are precisely what you want (or probably just one in the case of dating).

We succeeded in raising $16 million in this second round of financing for CombinatoRx. We were able to raise successively larger amounts of money over time, and in the end we raised over $350 million and employed more than 170 people. We found many striking examples of synergistic and unexpected combinations of approved drugs. However, in almost all of the cases there was something in the combination that didn't quite work in terms of turning it into a drug. Perhaps one of the drugs had an unacceptable side effect or wouldn't distribute to the right tissue for the disease we wanted to target. Or sometimes the two drugs remained in the animal or patient for different lengths of time (i.e., minutes versus days), or had incompatible chemical features. For a variety of reasons, many of the most striking combinations in cells and animals could not actually be turned into marketable products.

Still, there were several drug combinations that we confirmed to be effective in mouse models, and a few of these were shown to have a beneficial effect in patients. However, in these cases, the benefit was too small to warrant creating a marketed drug. Still, we remained hopeful that eventually the company would find the right combination that could be turned into a marketed drug.

THE FATE OF COMBINATORX

In 2007 I started to become less involved in the company. Partly this was due to the departure of a number of people I had most closely interacted with, including Curtis, who had served as Vice President of Research since the company's founding but then moved to Harvard to run a research fund. My reduced involvement was also partly because I found I no longer had much influence over the approach the company was taking. Despite my frustration, I considered this a natural part of the life cycle of a company, whereby a successful company eventually outgrows its founders and sets its own course. In time I found that I was hearing about the latest advances at CombinatoRx through press releases, like everyone else.

It was thus a surprise that at 8:30 AM on October 6, 2008 I read, along with the rest of the world, that CombinatoRx's main drug, Synavive, had reported disappointing results in an advanced clinical trial. I was discouraged by this, but I didn't realize the seriousness of the event. After all, the company had a pipeline of many drugs in preclinical and clinical development—this was just one drug candidate. Nonetheless, investors were spooked, perhaps because of the terrible economic and investment climate in the country at that time.

The stock price collapsed to under $1, meaning that the options held by the executives of the company were worthless, and that it would also be impossible to raise additional funds to support research at the depressed stock price. This meant the company had to hunker down, conserve cash, and last for some time without any additional funding. To survive, they were unfortunately forced to lay off 65% of their workforce relatively quickly, and then announced additional layoffs over time.

On July 1, 2009, in order to survive, CombinatoRx announced a merger with Neuromed to obtain an advanced clinical candidate that Neuromed was developing. As part of the merger Alexis resigned as CEO and handed over the company to new managers. Within a short time Neuromed's drug was approved for use by the FDA, representing CombinatoRx's first marketed drug, although the drug did not come from the combination screening approach.

CombinatoRx, renamed Zalicus after the merger, has not yet succeeded in creating a blockbuster drug by combining existing drugs in new ways. Nonetheless, what did we learn about combination drugs? First, we learned that the space of combination activities is heterogeneous (i.e., that the probability of finding a synergistic interaction between two drugs depends a lot on the nature of the two drugs).[9]

If the probability of any two drugs showing synergy were to be fairly constant, independent of anything particular about the two drugs, then the surface of an activity plot showing synergy would look like a thinly populated forest (see Figure 4.1). The height of a tree would tell you the amount of synergy of a given drug combination, and the spacing from one tree to the next would tell you about the frequency of these synergistic effects. Most crucially, we would expect the trees to be evenly distributed around the surface.

Instead, what we actually found was that the trees were more densely packed together in places where the two individual drugs were already active on their own. In contrast, the places where two drugs were inactive on their own looked like a barren plain, with hardly any trees (i.e., synergies) present at all.

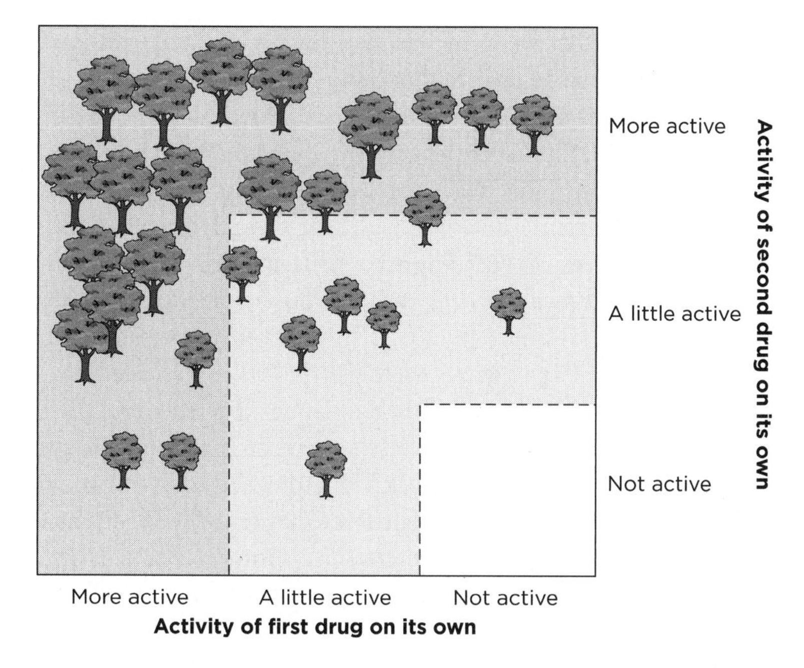

Figure 4.1 Combination activity space. The probability of finding synergy between two drugs depends on the level of activity of each individual drug. When each drug on its own is inactive, there is little to no chance of finding synergy between the two drugs. When one or both drugs is active on its own, the chance of finding synergy between the two drugs becomes increasingly large.

In the early days of starting CombinatoRx we dreamed of finding a combination in which two drugs were entirely inactive on their own, but when tested in combination, would have a powerful synergy. This kind of 0 + 0 = 100 synergy would be the most striking, the most unexpected, and therefore the most exciting. Also, such a combination would be the most patentable, precisely because it is so unexpected.

However, after testing millions upon millions of combinations, we never found a single example of this kind of synergy from nothing.

Instead, we saw that when two drugs were already active on their own, there was a good chance that they would have synergistic activities when tested together. In some cases the chance of synergy was almost 1 in 4. The lower the activity of a single drug on its own, the lower the chance that it would synergize with anything. When a drug by itself was totally inactive, then there was only a small chance that it would synergize with any other drug, and it was almost impossible to find an inactive drug that would synergize with another inactive drug.

This map of synergy space is useful for the future, because it suggests that you don't need to search blindly for unexpected synergies—instead you can focus your efforts testing new combinations of drugs that are already active by themselves. This is probably what you would have guessed intuitively, but it is valuable to have run the experiment and confirmed this intuition. Moreover, this information could be valuable in optimizing the treatment of patients who are taking multiple drugs at once.

My conclusion from the experience of working with such massive numbers of drug combinations is that one can find combinations that are synergistic and more effective than single drugs. On the other hand, it is quite challenging to turn these observations in the lab into something that can clinically benefit patients—and moreover, into commercial products.

Mainly, I have concluded that to solve the drug discovery shortage, we will ultimately need to search elsewhere. We will need to return to the idea of targeting undruggable proteins and unlocking the potential of the entire set of human gene products. We will need to find a way to create drugs that can bind to these challenging proteins. In short, we will need to drug the undruggable.

5

THE UNDRUGGABLE RAS PROTEIN

In recent decades, researchers have learned a great deal about the molecular causes of cancer and other diseases. The challenge facing the drug discovery community today is how to turn this wealth of information about basic molecular mechanisms—such as those governing tumor formation—into new medicines. A similar story can be told in many diseases—great strides have been made in revealing the genes that cause disease, but the largely unmet challenge is to create effective medicines based on this molecular information.

To understand the monumental difficulty of this challenge, it is instructive to examine the history associated with a specific protein—the RAS oncoprotein. This protein is a crucial contributor to the genesis of human cancers. In many respects it should represent an ideal starting point for creating a powerful, transformative, anti-cancer drug. The question, however, is whether the RAS protein is druggable.

THE CELLULAR NATURE OF CANCER

The discovery of the role of the RAS protein in cancer was preceded by extensive work investigating the fundamental question of how cancer arises; specifically, how cancer cells emerge from normal cells. These efforts have their origins in the early nineteenth century, when the cell theory was developed. In 1838–1839 Matthias Jakob Schleiden, a German botanist, and Theodor Schwann, a German physiologist, proposed that the basic building block of multicellular organisms is the cell; this defined the fundamental unit of living systems.[1]

According to this theory, plants and animals are composed of masses of cells, each specialized for different functions. Moreover, new cells are produced from preexisting cells, as opposed to being produced by spontaneous generation. Rudolf Virchow, another German scientist, expanded upon this theory by creating the field of cellular pathology, studying what goes wrong in cells and how these events are connected to disease. In this context, Virchow proposed that cancer cells arise from normal cells, based on observations of the properties of cancer cells in patients.[2]

This process by which normal cells might be converted into cancer cells was mysterious. How could a normal cell be transformed into such a different type of cell, one that could generate human cancers? The first step in solving this mystery emerged from the study not of tumor cells, but of bacterial cells. In 1928 Frederick Griffith, a British scientist, reported the results of an experiment in which the characteristics of one bacterial cell could be transferred to another bacterial cell. Griffith was working with two different strains of a bacterium called *Streptococcus pneumoniae* in an attempt to make a vaccine for the Spanish flu pandemic. Although the flu pandemic was caused by a virus (a particularly virulent strain of the influenza virus), many flu patients became superinfected by this bacterium, leading to a deadly pneumonia. Preventing infection of patients by *S. pneumoniae* might make it possible to lessen the damage caused by a future flu pandemic.

One strain of these bacteria appeared smooth under the microscope, and one appeared rough; only the smooth strain could cause pneumonia when delivered to mice. Thus the appearance (smooth or rough) of bacterial cells was correlated with their infectivity, or virulence. Griffith found that if he heated the smooth, infectious strain, it lost its infectious quality—the heating process destroyed proteins needed for infection.

In the crucial experiment, when Griffith mixed the heat-destroyed smooth bacteria (which were not infectious) with the live, rough bacteria (which were also not infectious), he obtained a strain that *was* infectious. The infectious nature of the dead smooth strain had been transferred into the living rough strain. Griffith had discovered that heritable changes could be transferred between cells. Presumably, a special type of molecule had been transferred from one cell type to the other; this remarkable molecule had the capacity to permanently change the infectious nature of the recipient cells, as well as all of their descendants.

Griffith didn't know the identity of the molecule that was responsible for transferring the characteristics of one cell to another.[3] In 1944 three American scientists identified this key molecule. Oswald Avery, Colin MacLeod, and Maclyn McCarty at Rockefeller University discovered that transmitting heritable information from one cell to another involved the transmission of a type of nucleic acid, called DNA, into the receiving cell.[4] Thus DNA was shown to be the molecule that contained heritable information.

TUMOR VIROLOGY AND CANCER-CAUSING GENES

This process of altering the properties of a cell by delivering DNA turned out to be crucial to the study of cancer. The introduction of foreign DNA into normal cells growing in the laboratory has been extensively used to discover cancer-causing genes. Cancer cells behave quite differently from normal cells—they grow without limit in the laboratory and are able to form tumors in some mice. Normal cells do not have these characteristics. Changing the DNA composition of normal cells can convert them into tumor cells.

This ability of DNA to convert normal cells into tumor cells was first revealed in the study of tumor viruses. A small number of cancers are caused by viral infections. The study of these cancer-causing viruses provided a foothold to identify the genes that are responsible for converting a normal cell into a tumor cell. Peyton Rous, at Rockefeller University, showed in 1911 that specific kinds of tumors could be induced via viral infection, founding this new field of tumor virology. In fact, most cancers are not caused by viruses, but the study of tumor viruses would in time lead to the identification of the core cancer-causing genes, which are involved in many kinds of cancers, even those that are not caused by viruses. The study of tumor viruses would lead to the discovery of the RAS oncogene and reveal its critical importance in many cancers. Thus the discovery by Rous that some tumors are caused by viral infection led to a series of experiments that would solve the mystery of how tumor cells are created from normal cells.

Despite the seminal importance of Rous's discovery, it was not appreciated by the scientific community. Rous was unable to convince other scientists of the importance of his findings. There were a series of technical and philosophical objections to his conclusions that gradually demoralized Rous. Within several years of his groundbreaking work Rous abandoned his approach. Perhaps because of this, the field languished to some degree. In time, however, the importance of his early work was recognized, and Rous went on to win a Nobel Prize in 1966 for his studies on Rous sarcoma virus.[5]

There is an interesting science history question here. Should we be concerned that other scientists did not appreciate the importance of Rous's discovery? Such concern is a common theme when reviewing the history of scientific discoveries—that a brilliant insight languished because of flawed common wisdom. However, any practicing scientist can tell you that rejection is truly the nature of science, more so than any other field. Many experiments fail to give the desired result, if they even work at all. Your manuscripts describing successful experiments are almost always rejected initially by reviewers and journal editors; only after multiple revisions and resubmissions do most studies see the light of day. Even Nobel Prize–winning scientists continue to suffer rejection of manuscripts and grant applications. This scientific process is designed to require the highest standards of evidence to introduce new results and ideas into the scientific literature. The greater the impact of a new idea, the higher the standard for the data supporting the idea.

The failure of Rous's contemporaries, as well as those of Farber and others, to accept breakthrough findings is what we would expect, and even hope for, in the scientific community. Most pioneering researchers are able to bear these rejections, albeit with grumbling. The most successful researchers become energized by rejection and are motivated to redouble their efforts to disprove their critics. Rous himself might bear some responsibility for failing to try harder to win over his critics with overwhelming evidence. This is a lesson for students as they enter a career in science. Do not be overwhelmed by rejection; instead, use it as a motivating force.

THE DISCOVERY OF THE GENE THAT CAUSES CHICKEN SARCOMAS

The virus studied by Rous caused connective-tissue tumors, known as *sarcomas*, to form in chickens. Renato Dulbecco, Harry Rubin, and Howard Temin discovered that connective tissue cells called fibroblasts that were grown in the laboratory after infection with the Rous sarcoma virus continued growing indefinitely, like cancer cells isolated from human tumors that were grown in plastic dishes in the laboratory.[6] The shape of the normal cells changed from flattened to rounded after viral infection; these infected cells also grew on top of one another and formed a small cluster of cells called a colony, and could grow in three dimensions, in a semisolid material; normal cells could not survive in such an environment. This process of converting normal cells into tumor cells was called *transformation*. These transformed cells did not need the normal growth stimulants that nontransformed cells required in order to divide. Cells transformed into tumor-like cells in the laboratory thus closely resembled cancer cells obtained from patients. This discovery that transformation, the conversion of normal cells into tumor cells, could be observed in a plastic dish in the laboratory opened up the possibility that the mechanism by which transformation occurs could be determined. Studying transformation using cell culture methods was much easier than using either patients' tumors or mice, which are quite a bit more complex to manipulate.[7]

Further studies on the mechanism of transformation by a number of investigators revealed a crucial feature of the virus-induced transformation process—DNA representing a single gene could transform normal cells into tumor cells.[8] The Rous sarcoma virus harbors four genes; however, not all four genes were required for the virus to reproduce itself in cells. Moreover, not all four genes were required for transforming normal cells into cancer cells. In fact, the genes needed to propagate the virus could be separated from the genes needed to transform normal cells into cancer cells, by making viruses with a modified set of genes. One such modification eliminated the ability of the modified virus to transform cells, without affecting the ability of the modified virus to replicate itself in cells. It was not the general process of viral infection, but one specific gene carried by the virus that caused transformation.

A similar line of experiments revealed that the protein produced by this viral "transformation gene" was needed to maintain cells in their cancer-like transformed state. That is, if this protein were incapacitated by mutation, the cells would revert back to normal, losing their transformed state. In principle, it might have been the case that the viral protein was only needed to start the transformation process; once the process was under way, the viral protein might be dispensable. In other words, once the cells were transformed, it wasn't clear that the viral protein would still be needed to keep the cells in this new state if transformation were an irreversible process. However, the experiments showed that in fact this viral protein was needed for both the initiation and the continuous maintenance of cell transformation, at least in this assay. What a powerful and interesting protein this must be.

In 1976 Harold Varmus and Mike Bishop, along with their colleagues Dominique Stehelin and Peter Vogt, discovered an unexpected and striking property of this viral sarcoma gene needed for transformation (dubbed v-SRC, pronounced vee-SARK, for its origin in the sarcoma virus). They found that this viral sarcoma-causing gene was similar in DNA sequence to a gene found in normal uninfected cells. They named this normal cellular gene c-SRC, for its cellular origin and similarity to v-SRC.[9] It had been thought that tumor viruses delivered unique viral genes into normal cells and that these viral genes were unrelated to the normal cellular genes present in uninfected cells. However, Bishop, Varmus, and colleagues demonstrated that normal uninfected chicken cells contained a gene that was closely related to the viral v-SRC gene, which suggested that the virus had copied the normal cellular c-SRC gene for the purpose of transformation. This caused a radical shift in the perceived nature of cancer—suddenly it was clear that normal cells could be transformed into cancer cells by delivering into them a suitably modified cellular gene. Thus cancer came to be seen as a disease of dysfunctional genes.

The normal chicken c-SRC gene was closely related in DNA sequence to the viral, cancer-causing v-SRC gene. However, the normal chicken c-SRC gene did not cause tumor formation but rather assisted in the normal development and maintenance of the chicken. Therefore, there must be some important change in the DNA sequence of the viral v-SRC gene that dramatically changed its effect on cells. The viral v-SRC gene was an oncogene—a cancer-causing gene. The normal cellular counterpart, the c-SRC gene, was capable of being converted into an oncogene and was therefore dubbed a proto-oncogene. This was the first of a string of discoveries showing that some genes within an organism have the potential, when suitably altered, to transform into genes with different effects on cells—to become cancer-causing oncogenes.[10]

In addition, there was a profound implication in this perspective. Although many tumor viruses have been identified, it is still true that the majority of cancers are not caused by viruses. The field of tumor virology struggled for years to gain the respect of the broader cancer biology community, because many researchers believed that tumor virology was an unusual exception to the causes of cancer.

Since most cancers are not caused by viruses, the mechanisms governing rare virus-induced tumor formation might be different from the common mechanisms governing tumor formation. From this perspective, studying the mechanisms of virus-induced tumor formation was studying a rare phenomenon, rather than the major cause of cancer.

The results of Bishop and Varmus, on the other hand, suggested that cancer could be caused by changes to normal cellular genes, in other words, to proto-oncogenes. Some viruses, such as the Rous sarcoma virus, appear to have captured these proto-oncogenes in an altered form, converting them into oncogenes.

THE DISCOVERY OF THE RAS ONCOGENE

Other tumor viruses revealed the importance of additional cellular proto-oncogenes that could be captured, activated, and used to cause tumor formation. The RAS oncogene was discovered in this way, using the Harvey and Kirsten sarcoma viruses. These two viruses contained aberrant versions of the RAS gene, so named for its role in causing rat sarcoma formation. Just as in the case of SRC, it was discovered in 1981 that the viral RAS (v-RAS) gene had a normal cellular counterpart, c-RAS. Again, c-RAS does not cause tumors, but v-RAS does. Thus c-RAS is a proto-oncogene and v-RAS is a bona fide cancer-causing oncogene.[11]

Does the cellular c-RAS proto-oncogene become activated in human tumors that are not caused by viruses? In 1982 three research groups (those headed by Robert Weinberg, Geoff Cooper, and Mariano Barbacid) discovered that the cellular RAS gene could indeed be mutated and activated in bladder, colon, and lung cancers.[12] In addition, RAS gene mutations were found in other models of cancer, such as models in which cancer-causing chemicals were applied to mice and tumors developed as a result. It was ultimately revealed that most cancers are caused by conversion of proto-oncogenes into oncogenes. Viruses are capable of effecting this conversion, as are

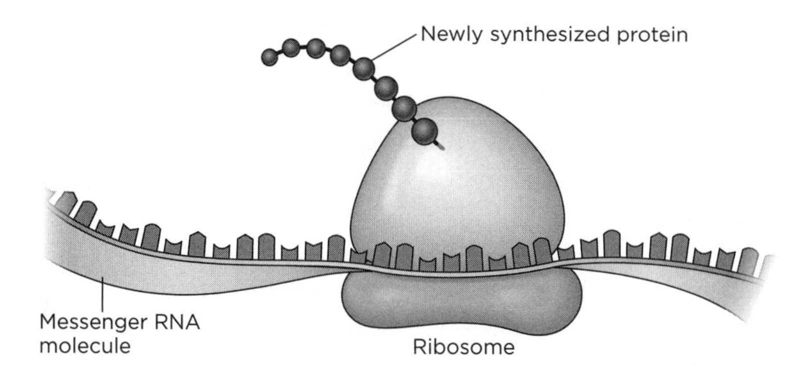

Figure 5.1 Translation of messenger RNA into protein. Messenger RNA molecules are converted into protein molecules by the ribosome, in a process known as translation.

cancer-causing chemicals called carcinogens. Tumor viruses provided the initial handle that ultimately allowed researchers to gain insight into one of the general mechanisms of tumor formation in all cancers, this mechanism being the conversion of proto-oncogenes into oncogenes.

To understand how mutations in proto-oncogenes can cause tumors to form, we must examine the so-called central dogma of molecular biology. DNA makes RNA, which makes protein. DNA and RNA are composed of repeating units of nucleic acid bases. DNA, representing the complete collection of genes in each cell, is transcribed into messenger RNA (mRNA) molecules, which represent each gene. These mRNA molecules are shuttled onto large protein-synthesis factories called *ribosomes*. In the ribosome, a sophisticated procedure enables each mRNA to be read and translated into a corresponding protein sequence with high fidelity (see Figure 5.1). Each group of three mRNA bases, called a *codon*, precisely determines the amino acid that will occur in the corresponding position of the protein, according to a table known as the genetic code.

For example, the first codon at the start of each protein-coding sequence is ATG, which encodes the amino acid methionine. Therefore, newly created proteins start with the amino acid methionine in their first position. The second amino acid is determined by the next three bases in the mRNA sequence, and so forth. When a change is made in the DNA sequence of a gene, the corresponding mRNA sequence is changed and then translated differently, causing a different amino acid to be inserted in the corresponding position of the protein.

Normally we think of mutations as disrupting the function of a protein by changing a specific amino acid in the protein, so it was surprising to see that mutations could activate a normally benign gene, turning it into a cancer-causing gene. Moreover, these activating mutations were extremely specific. Altering the identity of a single position in the DNA sequence caused just one amino acid to be altered in a string of 189 amino acids in the RAS protein. Nonetheless, this slight change in the amino acid makeup of the RAS protein caused a dramatic change in its properties and in the cells that harbor it. How could such an effect occur?

FROM RAS MUTATIONS TO TUMOR FORMATION

To answer this mystery, it was necessary to dig into the biochemistry of the RAS protein and determine exactly what the protein does in both normal cells and cancer cells. Indeed, RAS appears to be an undruggable protein, so understanding the molecular function of RAS reveals why some proteins appear undruggable.

Initially, Ed Scolnick and his colleagues found that the RAS protein could bind to nucleotides, a specific small chemical found inside cells.[13] Nucleotides are the starting materials used to create RNA and DNA, where they are strung together as billions of building blocks in sequence, encoding the genetic information of cells and organisms. Isolated nucleotides also serve other biochemical functions in cells. For example, ATP (adenosine triphosphate) is one of the four nucleotides that is used to make RNA (the nucleotide abbreviated as simply A) and is similar to a corresponding nucleotide that is used to make DNA. ATP also exists as an isolated molecule where it functions as the energy currency of the cell, in addition to serving as the phosphoryl donor for kinases, as discussed in Chapter 1. When food is broken down, the energy extracted is used to create ATP. When work in the cell needs to be performed, ATP functions like a battery, releasing the energy that was previously stored in this molecule.

Another nucleotide that serves a second function, in addition to its role in forming RNA, is GTP (guanosine triphosphate; known simply as G in the context of an RNA or DNA sequence). GTP is used to regulate the function of a large class of proteins, known as G proteins for their ability to interact with GTP. Several groups discovered that RAS is a G protein, meaning that RAS is able to bind to GTP. Moreover, the RAS gene has a similar DNA sequence to other genes encoding G proteins; this implies that the RAS protein has a similar amino acid composition to other G proteins and functions similarly to other members of this protein family. G proteins are frequently used as a means of transmitting information from one location in the cell to another, and RAS, being no exception, functions in this way.

GTP binds to RAS and activates its pro-growth signal, causing cell proliferation (see Figure 5.2). However, RAS possesses an intrinsic ability to convert GTP (guanosine triphosphate) into the related chemical GDP (guanosine diphosphate). In other words, when GDP is bound to RAS, which happens every time RAS catalyzes this reaction, the signaling function of RAS is shut off. The signaling function remains off until the GDP nucleotide can dissociate from RAS. Since GTP is normally much more abundant than GDP, and since RAS can bind both GDP and GTP well, when a new nucleotide binds to the empty RAS protein, it is likely to be a GTP nucleotide (as opposed to GDP), which reactivates the signaling function of RAS.

Thus RAS has an intrinsic off switch built deeply into its molecular function. GTP binds and activates RAS, causing the cell to divide and grow. After a short time, RAS converts GTP to GDP, shutting off signaling and blocking cell proliferation. A little while later the GDP dissociates and is replaced by a new GTP molecule that binds to RAS. The signaling activity is turned on and the cycle begins anew. It is like a light switch with a short timer on it, causing the light to

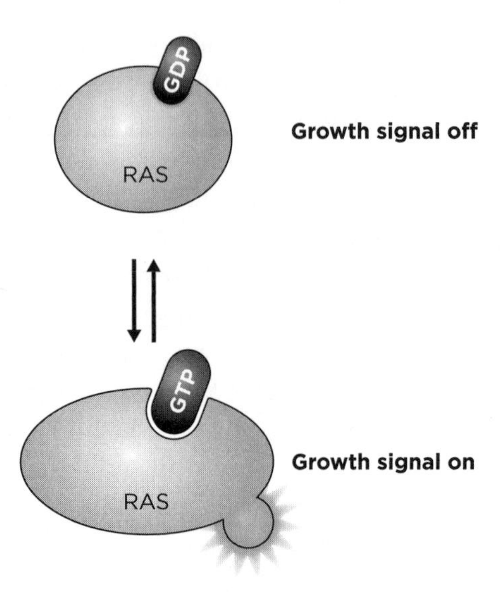

Figure 5.2 The RAS on-off cycle. The RAS protein can cycle between GTP-bound and GDP-bound forms. When GTP is bound, the signaling function of RAS is on and cell proliferation ensues. When GDP is bound, the signaling function of RAS is off and cell proliferation is prevented.

turn off automatically after a few minutes of activity. It is an ingenious little machine that evolution has created. However, it has a dark side.

The problem with this design of RAS is that it is easy to break in a way that leaves the light always on. A mutation that prevents RAS from converting GTP to GDP will cause RAS to be stuck in the "on" signaling position. This is precisely what cancer-causing mutations in RAS do. They prevent the ability of RAS to convert GTP to GDP, leaving the protein in an active signaling state permanently. Growth and proliferation are continuously signaled, causing unregulated growth, and eventually tumor formation. This is how a specific mutation can cause such a dramatic change. The mutation is causing a loss of function of the protein—it is easier to break a protein by mutation than to create a new function by mutation. However, it turns out, unfortunately, that RAS is designed so that when the biochemical function is broken, the signaling function is always on, causing tumor formation.

This unusual feature of RAS biochemistry suggests that creating a small molecule drug that targets RAS would be extremely challenging. Most drugs act by binding to a target protein and inhibiting the biochemical function of the protein. However, in the case of RAS, the mutant, oncogenic protein already has compromised GTP-to-GDP-converting activity. Therefore, a drug that binds to the normal c-RAS protein and inhibits its GTP-to-GDP-converting activity would act like an oncogenic RAS mutation—such a drug would actually cause cancer.

Another potential strategy would be to find a small molecule that can compete with GTP for binding to RAS. If a drug could prevent GTP (or GDP) from binding to RAS, then the signaling function of RAS would not be turned on, and tumors would not form. However, GTP binds extremely tightly to RAS. Its binding has been estimated to be about a thousand times better than the best drugs that have been developed. Therefore, it seems unlikely one could ever create a drug that binds tighter than GTP to RAS, in order to displace GTP and allow the drug to bind in its place. These aspects of RAS increasingly make this protein appear undruggable.

BLOCKING THE MEMBRANE TARGETING OF RAS

In order to create a drug that targets the RAS protein and blocks its cancer-causing activity, one must find a way to inhibit signaling by the protein, which is the process through which RAS transmits its cancer-causing message. A key advance in this strategy came with the discovery that the RAS protein is modified after it is synthesized in cells.

Once messenger RNA molecules are translated into proteins by the ribosome, the resulting proteins can be chemically modified in a number of ways. Such alterations to proteins are called *post-translational modifications*. These modifications include phosphorylation, the addition of a phosphoryl group onto a protein, which is the reaction carried out by kinases such as BCR-ABL, as well as other modifications. Each modified protein can behave quite differently, even though the amino acid sequence of the modified proteins is the same as the unmodified protein.

RAS is one of the proteins that is post-translationally modified. This discovery suggested the possibility that although the RAS protein itself appears to be undruggable, it might be possible to prevent post-translational modification of RAS, and thereby halt the cancer-inducing effects of mutant RAS. This discovery emerged in 1982, when it was found that RAS becomes attached to the outer membrane of cells. Shortly thereafter, it was found that a greasy lipid molecule, known as a farnesyl (FAR-neh-seel) group, is attached to the end of the RAS protein. This modification is *hydrophobic*, or water fearing, meaning that it prefers to localize to lipid membranes rather than to a water solution in the interior of cells.

Once the RAS protein has been attached to a farnesyl group, it is drawn to the membrane because of the attraction of the greasy farnesyl group for the oily environment within the membrane. The attachment of this farnesyl group onto RAS is required for the oncogenic signaling activity of mutant RAS. In other words, RAS can signal its growth-promoting message only when it is found at the cell membrane. This was an important finding, because it suggested that a drug that prevented RAS from localizing to the plasma membrane would block the oncogenic effects of mutant RAS. Suddenly there was a strategy for blocking mutant RAS, and overcoming its unusual biochemical nature.

This discovery led to the suggestion that a small molecule drug could be created that would inhibit the enzyme responsible for adding a farnesyl group onto the RAS protein. This enzyme is, appropriately enough, known as RAS farnesyl transferase, because it transfers this farnesyl group onto the RAS protein. Thus a number of pharmaceutical companies, including Merck, Schering-Plough and Johnson & Johnson, began intensively searching for RAS farnesyl transferase inhibitors. In fact, my first research experience in a lab was looking for farnesyl transferase inhibitors. At the time, I had completed my second year as an undergraduate at Cornell University, and I participated in a summer internship at Schering-Plough Research Institute in Kenilworth, New Jersey. I was excited about plumbing the depths of pharmaceutical knowhow and scholarship, and being surrounded by experts in drug discovery.

However, I found that the research had a rote quality to it, and I didn't see great enthusiasm in many of the scientists. Up until this point, I assumed that I would end up working in the pharmaceutical industry. However, after my summer at Schering-Plough, I began to doubt whether this would be a fulfilling career for me. After I graduated from college, I spent another seven months working at Schering-Plough with a different group, and again I was unsatisfied by the environment. I found that the research was certainly a challenge, and the people were clearly bright. Yet, there was a spark that was missing. I didn't go to bed thinking about the work we were doing. No one seemed to be trying to change the world, and that was the kind of energy I was looking for. This experience is what set me on the path to forging a career in academia, but I subsequently learned that lab culture is critical to research success. Other groups in industry have hunger and enthusiasm for discovery, and many groups in academia do not. I've come to realize that one aspect of creating a productive lab is creating an environment where everyone feels they have a chance to discover something profound.

The team of researchers I was working with at Schering-Plough was a group of chemists focused on creating a small molecule drug that could inhibit RAS farnesyl transferase. I was involved in creating some compounds that might serve as such inhibitors. Ultimately, the compounds I was pursuing were not effective candidates. However, the company did pursue another compound that was more promising, which was eventually turned into a drug named lonafarnib (Sarasar). This compound, like other farnesyl transferase inhibitors, has been studied for its potential in a wide array of cancers, without yet seeing striking activity.

One complication that arose concerned the specific RAS protein studied. There are three highly related RAS genes in humans, named K-RAS, N-RAS, and H-RAS. The H-RAS and K-RAS genes were the first discovered, and were named based on the viruses used in their discovery (Harvey RAS and Kirsten RAS, respectively). All three RAS genes are similar in sequence and function, and all three are mutated in human cancers.

However, there are some differences that turn out to be crucial. While the H-RAS protein is modified with a farnesyl group, the K-RAS and N-RAS proteins become modified not only with a farnesyl group, but also a longer 20-carbon molecule. Thus blocking the farnesyl transfer reaction that occurs on K-RAS and N-RAS does not block their membrane localization, because they still get attached to the 20-carbon molecule, and this modification is sufficient to localize them to the cell membrane. As a result of this difference in post-translational modifications between different RAS proteins, the localization of H-RAS is changed when farnesyl transferase is inhibited, but the localization of K-RAS and N-RAS are unaffected, and therefore the oncogenic effects of only H-RAS are blocked. Thus farnesyl transferase inhibitors can block the cancer-causing effects of only mutant H-RAS.

Unfortunately, of the three RAS genes, H-RAS is the least significant, in terms of the likelihood of being mutated in human cancers. These three RAS genes are mutated at different rates in cancers, for an unknown reason. For example, K-RAS is mutated in 60% of pancreatic cancers, but H-RAS is never mutated in pancreatic cancers. K-RAS is mutated in 19% of lung cancers and H-RAS is not mutated in lung cancers at all.[14] Overall, K-RAS is the most common of these RAS genes to be mutated in cancers, with one in five cancers of all types having a K-RAS mutation. However, H-RAS is mutated at a much lower rate, and only in uncommon tumor types.

Thus, lamentably for cancer patients, of the three RAS genes, it is possible to inhibit the membrane localization and oncogenic activity of only H-RAS, but this is the least important of the RAS genes. Unfortunately, it was the first discovered, and the initial cell culture experiments and mouse tests used oncogenic H-RAS. Hence, when the first farnesyl transferase inhibitors were developed and tested in cells and mice, they looked fantastic, causing tumor regression with few side effects. When they were tested in humans, who have mostly K-RAS mutations in their tumors, they were much less effective.[15]

THE FUTURE OF RAS AND CANCER DRUG DISCOVERY

So what options are left? Could researchers create drugs that block addition of the second 20-carbon molecule at the same time that transfer of the farnesyl group is blocked? Unfortunately, this seems to be too severe of an intervention and causes severe side effects. Moreover, there are now data suggesting that some of the efficacy of farnesyl transferase inhibitors is due to inhibiting modification of other proteins, such as one called RHEB (a shortened version of the description RAS homolog enriched in brain).[16] If this is true, it suggests that one wouldn't want to target these drugs to tumors with mutant RAS, but rather mutant RHEB, although it is not clear this would be a common event. Much of the enthusiasm for developing these drug candidates has waned.

Due to the great attractiveness of the RAS proteins as drug targets, and their central importance in cancer, other strategies have been attempted in recent years. There have been efforts to make RNA molecules that can interfere with the RAS messenger RNA, but such approaches have suffered from the central problem of large molecule drugs. Large molecules are difficult to use as drugs, because they are often unstable and don't penetrate into cells. Many small molecules don't suffer from these limitations, which is why the bulk of drug discovery involves creating small molecule drugs.

There has been some effort to target other proteins that act downstream of RAS. When RAS signals its tumor-promoting message, RAS must turn on other proteins to relay this message. One of these proteins is RAF; RAF is a kinase, which we now know to be a druggable class of proteins. In 1992 scientists at the company Onyx

Pharmaceuticals searched for a RAF inhibitor, and they found a good candidate. This compound was optimized, and the result was a drug that was named sorafenib.[17] In recent years this drug and its target protein RAF have been the subject of great interest, because RAF is itself mutated in a number of cancers.

When drugs such as sorafenib are ready for testing in human patients, there is a rigorous, regimented three-stage process for assessing their safety and efficacy. In phase I the drug is tested mostly for safety, usually in a small number of patients at a series of doses to determine what dose level can safely be given to patients. In phase II the drug is tested in a few hundred patients for both safety and efficacy; it is critical that the drug is proven effective in this stage. Finally, in phase III, the drug is tested in thousands of patients to unequivocally demonstrate both safety and efficacy in a large population.

When sorafenib was tested in the initial phase I human cancer trial, the researchers included all tumor types, irrespective of their RAF or RAS gene mutation status. Surprisingly, a significant effect was seen mostly in renal cell (kidney) cancers. Although sorafenib was discovered and developed as an inhibitor of RAF, because of the similarity of many kinases, it also inhibits several other kinases, including the platelet-derived growth factor, the vascular endothelial growth factor (VEGF) receptors 2 and 3, and the c-KIT receptor. Sorafenib's beneficial effect in patients with renal cell carcinoma seems likely to be due to its activity against the VEGF receptors, rather than against RAF. So although sorafenib seems to be a promising drug for renal cell carcinoma, the challenge of tackling mutant RAS tumors remains unmet.

We have come a long way since the hypothesis of the cell theory in 1838–1839 by Schleiden and Schwann. We have learned that cancer is caused by mutations in specific genes. We have learned an enormous amount about hundreds of cancer-causing genes, in terms of their normal biochemical functions, their interaction partners in cells, and how and when they are mutated in human tumors. Yet, for all the elegance and beauty of this intellectual edifice, we have been stymied by the challenge of using our knowledge to create effective drugs targeting most of these molecular alterations, imatinib and BCR-ABL being one of the few exceptions. Many cancer drugs were created in the golden era of cancer chemotherapy, and are most effective against rapidly dividing cells. Some additional strategies have been explored, with limited success, such as angiogenesis inhibition and targeting hormone-dependent tumors.

However, to break through to the next frontier in cancer drug discovery, we need to solve the problem of targeted therapy. This approach involves discovering and then inhibiting the specific proteins that are mutated in cancers, that drive tumor formation, and that are needed for continued maintenance of the transformed state. There have been some hints of the potential of this approach, as evidenced by the discovery and utility of imatinib and a small number of other drugs.

Yet, we have failed at the big challenges, like tackling mutant RAS. We have known about the important role of RAS in cancer for almost three decades. Nearly one in five tumors contains a mutation in one of the RAS genes. Despite the great efforts expended, and the central importance of RAS, we have failed to create a drug that eliminates RAS mutant tumors. We cannot do the seemingly impossible—we cannot inhibit RAS directly, and we cannot block the oncogenic function of RAS in a specific and effective way. Most researchers have come to the conclusion that RAS cannot be targeted, at least not directly, and they have moved on to other research problems.

This is the great challenge we face. With the limited size of the genome, it is not clear that we can afford to walk away from these challenges, as eventually we will run out of tractable targets. On the other hand, we just can't seem to make progress against RAS, and other similarly challenging proteins.

The same story can be told for hundreds of other crucially important disease-causing proteins. We know of dozens of cancer-causing genes, the vast majority of which are refractory to targeting, similar to RAS. We know of genetic causes of many neurodegenerative diseases, such as Alzheimer's disease, Huntington's disease, amyotrophic lateral sclerosis, Parkinson's disease, spinal muscular atrophy, and many others. Researchers continue to assemble data and knowledge about the biochemical and biological functions of these disease genes and their interacting partners. Yet, the more we learn, the more challenging these problems appear, for nowhere in sight is there a druggable target that could cure any of these diseases.

In the face of such data, some may despair. Some may reasonably conclude that despite the depth of knowledge we have accumulated, we are reaching the limits of therapeutics. In this view, many of these diseases may be inherently incurable, and we may need to lower our expectations in this regard. We may need to accept the inevitability of cancer, neurodegeneration, and other ailments. We may have come to the end of productive drug discovery.

Or perhaps not. It is possible that there are solutions, lurking at the periphery of our current knowledge. Will some unorthodox approach being attempted today revolutionize our thinking tomorrow? To answer this question, we must explore what approaches can be taken to solve this potentially devastating drug discovery roadblock. We must take a broader view of the genome and try to understand which proteins are druggable and why. In this way we may break through the drug discovery bottleneck and finally be able to tackle these undruggable proteins—and the incurable diseases they control.

6

THE DRUGGABLE GENOME

The debate about the future of medicine concerns the nature of the druggable genome, which is the subset of the human genome that encodes proteins that can be affected by drugs. To understand what this encompasses, we need to understand the composition of the human genome—the complete collection of DNA found inside of each of our cells that contains all of the instructions needed to create a human being from a fertilized egg. What are all of these genes, and the proteins they specify, and which are capable of binding to drugs?

THE BIRTH OF GENETICS

The study of the human genome began with Gregor Mendel in 1853. Mendel was a devout monk and teacher who worked in a monastery in what is now the Czech Republic, beginning in 1843.[1] Stimulated by a lecture given by Franz Unger about the nature of heredity, Mendel performed experiments on pea matings.[2] He was interested in understanding the mechanisms governing the passing of characteristics from generation to generation. Clearly, traits are passed from parent to offspring in all species, and breeders had studied this process for centuries. Nevertheless, no one had uncovered the *mechanism* of transmitting traits from one generation to the next or found an explanation for why some offspring were like one parent or the other, and why some offspring were quite different from both parents.[3]

Folk wisdom regarding inheritance of characteristics suggested that the traits of parents should simply be blended in progeny. In other words, the mating of a tall person and a short person should produce medium-size children. For some traits, such blending is observed, but for others (like eye color) it is not. James F. Crow, a senior population genetics researcher at the University of Wisconsin at Madison, pointed out to me that a key aspect of the blending theory is that once two traits are blended, it is not possible to recover the original traits in a future generation. He likened this to blending paints—once you blend red and yellow paint to make orange paint, there is no going back to the original red and yellow colors. The consequence of this blending theory is that once two individuals mate, their differences would be irrevocably lost. Charles Darwin himself subscribed to this blending view, and it caused him to be seriously concerned with generating new variability to counter blending-induced sameness. If Darwin knew of Mendel's work, he wouldn't have been so concerned with this issue, because according to Mendel, traits can be readily recovered in future generations, as we shall see.

Mendel discovered that combining plants with different traits would yield offspring with a precise ratio of the unblended traits, such as a 3-to-1 ratio in the second generation. Mendel crossed pea plants with unusual characteristics and quantified the presence of these characteristics in the offspring. For example, he crossed plants having smooth seeds with plants having wrinkled seeds, or plants that flowered along the stem with plants that flowered at the end of the stem, and found this reproducible numerical relationship among the type of offspring produced in each generation.[4]

Mendel proposed a mechanism to explain this numerically precise trait inheritance. He suggested that traits such as whether seeds are wrinkled or smooth are determined by "factors" passed from parent to offspring. In his crucial insight, Mendel proposed that each parent has two versions, later named *alleles*, of each factor, which later came to be called *genes*, and that offspring get one allele from each parent.[5] If both parents have one allele *A* and one allele *a* (i.e., having *Aa* encoded in their DNA), then there are four possible products of these parents mating with each other: *AA*, *Aa*, *aA*, and *aa*. Think of *A* as brown eyes and *a* as blue eyes. If *A* (brown eyes) always dominates *a* (blue eyes), then three of these four possible offspring will look like *A* (i.e., they will have brown eyes) and only one of four will look like little *a* (blue eyes). The 3-to-1 ratio is determined by the intrinsic mechanism of the process of inheritance.[6] In 1866 Mendel published these monumental findings, but few recognized the importance of his work.

Mendel tried to get recognition from other scientists by sending reprints of his work, including one to Unger, but almost everyone ignored his findings.[7] Mendel died of kidney disease in 1884, largely unknown and himself unaware that he would eventually be recognized as the founder of the science of genetics.[8] However, in 1900, three botanists, Carl Correns, Erich von Tschermack, and Hugo de Vries, rediscovered Mendel's paper and vigorously disseminated his findings to the scientific community.[9]

While Mendel's research was being rediscovered by his new advocates in the early twentieth century, Thomas Hunt Morgan was working as a zoologist at Columbia University, in New York City. Morgan was an extremely sharp and prolific thinker. Throughout his career, his research touched on embryology, regeneration, sex determination, and evolution. In 1905, however, he became interested in this question of heredity.[10] Curiously, at this point in his career, Morgan rejected Mendel's theory of inheritance. He also rejected another theory of inheritance, based on chromosomes, advocated by his colleague Edmund Wilson also at Columbia. In 1909 Morgan was still attacking the theory of Mendelian inheritance. However, by 1910 he was able to create a synthesis of these two theories reversing his earlier objections and becoming a strong advocate for both theories.[11]

The chromosome theory of inheritance, which hypothesized that sex and other characteristics are determined by the inheritance of specific molecules within cells called chromosomes, was supported by Hermann Henking's discovery in 1891 of the accessory chromosome.[12] This was a Y-shaped chromosome that didn't have a counterpart, unlike all other chromosomes, which come in pairs. This curious chromosome could be found in males of a number of different species.

In 1901 Clarence Erwin McClung focused on this problem of chromosomes and sex determination. McClung was an American zoologist who grew up, personally and professionally, in Kansas.[13] His interest in science began as a young child, when he, like Justus von Liebig, was intrigued by chemistry. This led McClung to obtain several degrees including a graduate in pharmacy degree, bachelor's degree, master's degree, and doctorate from the University of Kansas. McClung studied a nuclear substance involved in sperm production in the long-horned grasshopper, *Xiphidium fasciatum*. In due time he showed that this unusual substance was a chromosome; moreover, he showed that grasshopper chromosome type correlated with the sex of the grasshopper. Grasshoppers don't have a Y chromosome, so the key distinction is the number of copies of the X chromosome.[14] McClung made the mistake of concluding that the X chromosome was male determining, because he used a mistaken count of chromosomes in the female provided by his student Walter Sutton.[15] This error was corrected in 1905 by Nettie Stevens, who studied a species of beetle and found that female beetles were XX and male beetles were XY.[16] Nonetheless, despite these results, it was difficult to prove that sex was directly determined by chromosomes.[17]

Morgan was opposed to the idea that sex was determined by chromosomes, because he didn't think linking an outcome like sex to a chemical structure provided a satisfying mechanistic explanation; in other words, it didn't explain *how* a chromosome determined sex. He favored the view that development consisted of taking undifferentiated cells (i.e., sexless cells) and turning them into differentiated adult tissue (i.e., containing a definite sex). He was skeptical of the chromosome theory because it implied that the character of sex was predetermined in the embryo, and therefore that development consisted simply of magnifying an intrinsic aspect that was already present in the earliest stage of life.[18]

Nonetheless, a crucial characteristic of the most insightful researchers—and scientists in general—is the ability to change their minds in the face of new data. In 1907 Richard Hertwig showed that the source of sperm used in frog fertilization experiments had a dramatic impact on the sex of the offspring, suggesting that sperm, and by extension chromosomes, really are sex-determining structures. Shortly thereafter, Morgan began breeding the fruit fly, *Drosophila melanogaster*, in his laboratory, with the goal of studying heredity through mutation. These flies are tiny creatures with small red eyes. For two years, he had little success at introducing mutations into these flies. He considered the entire project a waste of resources. However, early in 1910 he made an observation that would change his life and create the foundation for an experimental science of genetics.[19]

This crucial discovery began with his observation of a single fruit fly with white eyes instead of the normal red color. There has been much speculation about the source of this famous white-eyed fly. William Ernest Castle, who studied mammalian development and inheritance at Harvard, claimed to have sent several white-eyed mutants to Morgan, but Morgan and his colleagues insisted that these did not survive the trip. Frank Lutz, who was a devoted entomologist, also found a white-eyed fly and sent some of its descendants to Morgan, but these also failed to thrive. Morgan recalled that it was a newly occurring natural mutant that produced the white-eyed fly that he studied so carefully in his lab. In fact, when Morgan visited his wife in the hospital in early January, he reportedly "talked of nothing but his new mutant," implying that its origin was shortly before this time.[20]

This conclusion was confirmed by the recollections of James Crow in a recent lecture about the early years of *Drosophila* genetics. Crow also spoke about Morgan's fly lab, in which these seminal experiments were performed in close-knit quarters by Morgan and his students. Crow saw the fly room in 1947, when he was invited to visit Columbia by Theodosius Dobzhansky, a later colleague of Morgan's. Crow stayed in the same guest room that had earlier been used by J. B. S. Haldane, the renowned British geneticist and biologist. When

Haldane arrived in New York, he made a comment to the media about the difficulty of finding a suit in America that would fit him, and due to his fame, many companies began sending him tailored suits, shipped to his room in New York. Even many months later, these suits kept arriving while Crow was staying in the room.

Once Morgan had his mutant fly, he bred this white-eyed fly to a red-eyed fly and found the precise ratio of red to white eyes in offspring as predicted by Mendelian genetics, with one striking difference—the white-eyed flies were all male. This showed that sex and other characteristics, like eye color, could be linked during the process of heredity, which could be explained by their location on chromosomes. This observation transformed Morgan's thinking about the mechanism of inherited traits. He proceeded to embrace both Mendelian genetics and the chromosome theory of inheritance, and in fact synthesized these two theories to create the science of genetics, confirming that inheritance of chromosomes was the mechanism of Mendelian processes.[21]

Together with his students, Morgan argued that genes, which determine visible characteristics such as having white eyes, were linearly arrayed along chromosomes.[22] In other words, chromosomes contain genes, the units of heredity. Before Morgan, genes were a hypothetical entity; because of Morgan's work, genes became associated with chromosomes and hence concrete and specific entities within cells.[23] In time Morgan became the leading geneticist in the world, being recognized with the Nobel Prize in 1933.

Barbara McClintock, the 1983 Nobel Laureate, generated cytological data (i.e., images of cells) supporting the notion that genes are actually pieces of chromosomes, and not something else associated with chromosomes. McClintock was adept at generating beautiful images of chromosomes in action. When she was starting as a graduate student at Cornell, she began working with Lowell Randolph, who had been trying to visualize chromosomal disorders in which maize plants had three copies of one chromosome, instead of the normal two. Randolph had been trying the usual techniques for years, without success. Within three days of starting on this project,

McClintock developed two technical innovations that solved the problem and allowed beautiful images of maize chromosomes to be obtained.[24] McClintock was later able to take images of maize chromosomes that had undergone recombination, which demonstrated that the process of genetic recombination involved chromosomes. This further solidified the notion that genes are found on chromosomes, and that inheritance of chromosomes determines the observable characteristics in organisms.[25]

FROM DNA TO GENOMICS

To understand the chemical makeup of chromosomes and genes, researchers turned to nucleic acids, which were first identified by Friedrich Miescher in 1869.[26] There are two major types of nucleic acids in cells—ribonucleic acids and deoxyribonucleic acids, abbreviated as RNA and DNA. These nucleic acids are composed of repeating sequences of four building blocks. DNA specifically consists of repeating units containing the bases adenine (A), guanine (G), thymine (T), and cytosine (C); RNA is similar but uses the base uracil (U) in place of thymine. These repeating units form a polymer, which is similar to a protein. The building blocks are different in proteins and nucleic acids, but the concept is similar. Nucleic acids are built from repeated units containing these four bases, whereas proteins are built from repeating units of 20 possible amino acids.

DNA was found to be localized to the nucleus, and then specifically to chromosomes. In these early days, mitochondrial DNA was not detected or studied, but we now know that a small amount of DNA is also found inside mitochondria, the energy factory of the cell, where it codes for the synthesis of a small number of mitochondrial proteins.

There was considerable debate about the significance of the mysterious nuclear DNA substance. A series of experiments, including one by Alfred Hershey and Martha Chase in 1952, showed that DNA, but not protein, was capable of transmitting heritable changes in organisms.[27] Thus it was concluded that chromosomes are composed of DNA, and that DNA is the material that controls the inheritance of traits.

On April 26, 1953, Watson and Crick published the double helical structure of the most common form of DNA in *Nature*. This structure revealed that each G was paired with a C and each A was paired with a T. This atom-by-atom picture of DNA solved the question of how genes were copied and passed on from parent to offspring—the two strands of DNA separate from each other, and each strand contains the same information. Filling in a new second strand on each of these separated DNA strands allows for replication of DNA without loss of information.

Through nearly a century, researchers thus progressed from Mendel's first articulation of the principles of inheritance to the atomic-resolution structure of DNA. The linear sequence of DNA is transcribed into a messenger RNA (mRNA) molecule, which is then translated on a ribosome into a sequence of amino acids, forming a protein.[28] These translated proteins then carry out the work of the cell by catalyzing enzyme reactions, forming structural components of cells, and acting as secreted factors that engage other cells and cause them to change the patterns of expressed genes. This flow of information, from DNA to RNA to protein, is the central dogma of biology.

Knowing this, one might be tempted to ask, can we identify the exact DNA sequence of a gene, and therefore the exact amino acid sequence of the translated protein molecule? Indeed, in 1976 Allan Maxam and Walter Gilbert developed the ability to determine the precise base sequence of DNA. Around the same time, Frederick Sanger developed an alternative methodology that eventually became widely used. Sanger took a conceptual leap and asked whether he could use this new DNA sequencing technology to determine the base sequence of an entire genome.[29]

A *genome* is the complete set of genes of an organism. In terms of DNA, the genome refers to the complete DNA sequence of every chromosome of an organism. In 1977 Sanger reported the complete

DNA sequence of the genome of phiX174, a small phage virus that infects some bacteria, such as *Escherichia coli*.[30]

PhiX174 has 11 genes within 5,386 bases.[31] Sanger's work meant that for the first time, it was possible to determine the entire DNA sequence of an organism's genome, allowing a new level of analysis based on having complete knowledge of the genetic makeup of the organism. In other words, you could know not just some of the genes present, but the complete list of genes present. This implied that you could know exactly how many members of each gene family were present, and that you could know which families of genes were *not* present in the organism, for example, if they had been lost in the course of evolution. The geneticist Eric Lander has called the emergence of complete genome sequences, or *genomics* as the field is called, the geneticists' equivalent to the periodic table—both are complete compendiums, of the building blocks of matter on the one hand, and of life on the other.[32]

The advent of DNA sequencing also provided another key insight into biology. By looking at the DNA sequence for two different genes, it became possible to compare these sequences to see how similar the genes were. Frequently two genes have somewhat similar or even extremely similar DNA sequences. In other words, when you compare the pattern of C, G, T, and A bases that make up these two genes, you will see that these sequences are nearly identical in some parts. This implies that these two genes encode proteins that have similar shapes and properties.

With a complete genome sequence in hand, it is possible to group genes into families based on similarity of DNA sequences. This is powerful, because it allows researchers to infer the function of a gene that has never been studied before. For example, as we saw in Chapter 1, the BCR-ABL gene encodes a protein that functions as a kinase, which adds a phosphoryl group onto other proteins. If a new gene has a DNA sequence that is similar to the BCL-ABL gene, and particularly if this similarity occurs in the region of the BCR-ABL gene that encodes this kinase function, then we would suspect that the novel gene also encodes a kinase that is involved in adding a phosphoryl group onto some other protein. Similar DNA sequences encode proteins that perform similar biochemical functions.

Moreover, DNA sequencing has revealed that all life is encoded by genes with similar sequences. Kinases are found in humans, monkeys, mice, worms, flies, yeast, and bacteria. There is a conservation of biochemistry across all species. This is why simpler organisms such as yeast, worms, and flies serve as reasonably accurate models for many processes occurring in humans.[33]

It was not long after the complete genome sequencing of phiX174 that geneticists began to imagine complete genome sequencing for higher species, including humans. The goal of identifying the genes causing human diseases gave additional urgency to this vision. By having the complete human genome sequence, it was argued, it should be possible to identify many disease-causing mutations. In 1993 the gene responsible for causing Huntington's disease was identified using this approach of human genetics.[34] Moreover, it was thought that complete genome sequences of humans and related species might shed light on what the unique attributes are of humans, and why we are so different from closely related species.[35] It was with such notions in mind that the Human Genome Project was launched in 1990, with the goal of determining the 3 billion base sequence of the human genome.[36]

The project began with simpler model organisms having smaller genomes. The first bacterial genome, *Haemophilus influenzae* with 1.8 million bases, was completely sequenced in 1995. Then the 12 million bases of the model yeast *Saccharomyces cerevisiae* was completely sequenced in 1996. Genome sequences began falling in rapid succession. The nematode *Caenorhabditis elegans'* genome with 97 million bases was completed in 1998, and shortly thereafter, the genomes of the plant *Arabidopsis thaliana* and Morgan's fruit fly *Drosophila melanogaster* were sequenced.[37] Before long, all eyes were on the coveted human genome.

It is important to realize the unprecedented nature of the Human Genome Project. This was a project on a scale that had never been contemplated before in biology and with a goal that seemed absurd only 25 years ago—to decode the 3 billion DNA letters that make up the genetic instructions for creating a human being. The project required new technologies that didn't exist when it was launched and a scale of international collaboration and cooperation that had not occurred in biology. Although thousands of people contributed in crucial ways to the international public human genome project, Eric Lander was the key driver and manager who successfully brought it to fruition.

Lander grew up in New York City, and did his graduate studies at Oxford University. He began his academic career as a professor at Harvard Business School, but before long he was drawn to biology. Given his wizardry with numbers, he was able to develop innovative approaches to using DNA sequence information for identifying human disease genes, which motivated him to tackle the project of sequencing the human genome. He started his biology career by moving across town and becoming a Fellow at the Whitehead Institute at the Massachusetts Institute of Technology (MIT). When he arrived at the Whitehead Institute from Harvard Business School, he asked for a small room and one computer. A few years later, he was a full professor at MIT, a senior member of the Whitehead Institute, and Director of the Whitehead/MIT Center for Genome Research, supervising an annual budget in the tens of millions and a staff of hundreds. He is now Director of the Broad Institute, which is located in Kendall Square and affiliated with MIT and Harvard.

I remember the first time I saw Eric speak at a genomics conference in Miami, Florida. I was struck by his creativity and rigor, which were interwoven like fine threads through his seminar. When I worked at the Whitehead Institute as a fellow scientifically displaced traveler (he being a mathematician and me being a chemist, both of us immersed among biologists), Eric was kind enough to take me under his wing, giving me weekly advice and substantial funding to try new ideas. It was an invigorating experience to work with such a generous mentor.

On February 12, 2001, the draft human genome sequence was published by two groups—the large international public consortium led by Eric and the private company Celera, led by J. Craig Venter. The public sequence was refined over the next few years, resulting today in a freely available electronic encyclopedia of the complete human genome.[38] It was a major breakthrough in biology and genetics.

THE DRUGGABILITY GAP

The sense of complete information was thrilling for biologists and disease gene hunters. It finally became possible to look at the complete collection of human genes and their functions without missing any pieces. The genome sequence was highly anticipated, given that it was more than a decade from launch of the project to its completion, and indeed enabled a transformation of both basic biology and disease biology, because of this completeness to the catalog of human genes.

However, this completeness also had an unexpected darker side. For the first time, it became possible to calculate the number of possible drug targets, and therefore, in some sense, the number of future drugs. Prior to this, there was a frontier quality to drug discovery research in the sense that no one worried about running out of drug targets or drugs in the future. As with the closing of the frontier in the American West in the nineteenth century, the complete list of human genes brought a change in psychology for many drug discovery researchers, as they contemplated the point at which drug discovery would finally come to a halt—that is when all the possible protein targets for drugs would be mined.

Initially there was a sense of enthusiasm as researchers anticipated that hundreds or thousands of new drug targets would be discovered by the human genome project, vastly enriching the drug discovery enterprise. As early as 1996, Jürgen Drews of Hoffmann–La Roche suggested that the genome project might reveal 10 times as many drug targets as had ever been discovered.[39] He calculated that all known drugs that had ever been developed targeted only 417 proteins. He further suggested that there could be 3,000 to 10,000 new drug targets encoded in the human genome, meaning many new proteins remained to be discovered and exploited. However, he didn't discuss in depth the issue of druggability—to what extent would it be possible to discover small molecules that can bind to these thousands of new targets and modulate their functions in a way that ameliorates disease?

In 2002, shortly after the human genome sequence was made available, this optimism began to turn to anxiety for some. Andrew Hopkins and Colin Groom, of Pfizer, published an analysis of the druggable genome.[40] First, they updated the number of druggable proteins. Their analysis revealed 399 protein targets that had been shown to bind drug-like compounds with reasonable affinity, even if some of these compounds weren't actually drugs. It was thought that the ability of a protein to bind to a small molecule with reasonable affinity suggested that such a protein could likely be targeted at some point with a drug that could eventually be approved for clinical use in humans. In a subsequent study Hopkins and colleagues found 207 protein targets for marketed small molecule drugs.[41] So their analyses suggested that out of 20,500 proteins encoded in the human genome, there are between 200 and 400 known druggable proteins, depending on how stringently one defines the term druggable. This was the starting point for their analysis of the extent of the druggable genome.

The human genome sequence revealed that human genes fall into distinct families based on their DNA sequence and the presumed biochemical functions and three-dimensional structures of their predicted protein products. Hopkins and Groom asked which families the 399 known druggable proteins were derived from. The idea was that if one kinase, for example, had been shown to be druggable, then in principle any kinase should be druggable given a suitable effort. About half of the 399 druggable proteins fell into just five families: G-protein-coupled receptors (GPCRs), kinases, proteases, nuclear hormone receptors, and phosphodiesterases, with the remainder scattered through some additional protein families (see Figure 6.1).

These five families represent the major players in terms of drug targets and mostly function by sending messages from one place to

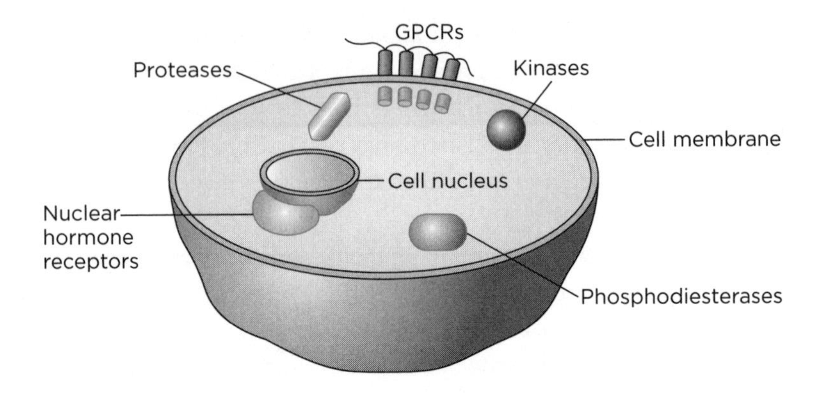

Figure 6.1 The most common protein targets of known drugs. The illustration shows the common types of proteins that are targeted by known drugs.

another in cells. Kinases are enzymes such as BCR-ABL that transfer a phosphoryl group from adenosine triphosphate (ATP) onto proteins; kinases usually act as signaling enzymes. Imatinib, used for treating chronic myelogenous leukemia, as discussed in Chapter 1, is an example of a drug targeting a kinase.

Nuclear hormone receptors are another class of signaling proteins. These proteins are found outside of the cell nucleus in their inactive state; when a hormone such as estrogen enters the cell and binds to a hormone receptor, it causes the receptor to move into the nucleus and turn on expression of new genes. These nuclear hormone receptors thus announce the presence of hormones to the rest of the cell. Anti-estrogens such as tamoxifen are examples of drugs that target nuclear hormone receptors—these are used to treat breast cancer.

Phosphodiesterases (FAHS-foh-dai-ESS-ter-ays-ez) are enzymes that break down a specific type of bond called a phosphodiester bond. These bonds are found in some signaling molecules; these enzymes act to shut off these signals. Viagra, used to treat impotence, is an example of a drug that acts by inhibiting a phosphodiesterase.

G-protein-coupled receptors are found on the surface of cells and act by sensing molecules outside of the cell and relaying a message into the cell about the presence of these extracellular molecules. Zantac, also known as ranitidine, is a drug used to treat acid reflux; the drug acts by binding to a GCPR known as the H2 histamine receptor, which is involved in regulating acid secretion in the stomach.

Finally, proteases are protein-destroying enzymes—they function by breaking down other protein molecules by cutting them at specific points along the protein chain. The HIV protease inhibitors are a class of drugs that prevent the replication of the human immunodeficiency virus, which causes acquired immunodeficiency syndrome (AIDS). These inhibitors act by blocking the function of the protease encoded in the viral genome, which is needed for viral replication.

Hopkins and Groom then asked how many genes out of the entire human genome encode proteins that could reasonably be thought of as druggable. They found 3,051 proteins to be within families where at least one member of the family has been shown to be already druggable. That represents 15% of proteins encoded in the genome. A subsequent analysis by Andreas Russ and Stefan Lampel in 2005, after the finished, high quality sequence of the human genome was published, found a similar number of potentially druggable proteins.[42]

There are three limitations to this estimate. First, it assumes that if one kinase, for example, has been targeted with a drug, then any kinase can be targeted with a drug. It is likely that there will be some members of each family that are resistant to targeting with small molecules, based on idiosyncrasies specific to these proteins; this means the 3,051 number may be an *overestimate*. Second, there are likely to be other families that haven't yet been targeted by small molecules but could be targeted in the future. We will look at some of these cases in future chapters. This type of error means the 3,051 may be an *underestimate*. Perhaps these two errors will cancel each other out to some degree.

Third, however, we haven't calculated what fraction of these druggable proteins are actually disease-modifying proteins, meaning that drugs targeting them would alter the outcome of a disease process.

This is a difficult number to calculate, but some estimates suggest that 10 to 15% of all proteins could be disease modifying. Thus there might be only between 300 and 450 disease-modifying, druggable proteins encoded in the entire human genome. If we have already targeted 399 proteins with drug-like small molecules, we may have identified all of the viable drug targets.

This is a grim conclusion, tentative as it may be. This is the basis for the pessimistic view of many scientists involved in drug discovery; there is a distinct possibility that we are running out of druggable protein targets. If this view is correct, pharmaceutical and biotechnology companies will have fewer and fewer new drugs, lower growth rates, and greater competition. Ultimately, we would witness the end of drug discovery, meaning that many ailments will go untreated. Intractable diseases, such as many cancers, Alzheimer's disease, and Parkinson's disease, will be uncured and will in fact become incurable, as we reach the limits of small molecule drug discovery.

However, there is a possible solution to this dilemma. Perhaps if we could zoom in and see what proteins look like up close, we could understand why some proteins are druggable and others appear undruggable. To do this, we would need, in effect, an extremely powerful magnifying glass—one that would allow us to see the individual atoms that make up proteins and how these atoms are arranged in each protein. Indeed, such a close-up view of proteins could unravel the deep mystery surrounding protein druggability, and might just perhaps enable researchers to create the next generation of medicines that target these seemingly intractable proteins.

PART TWO

THE PATH TO THE NEXT GENERATION OF MEDICINES

7

PEERING INSIDE PROTEINS

To understand why some proteins are more susceptible to modulation by small molecule drugs than other proteins, it is essential to zoom in to the atom-by-atom picture of proteins, and observe where and how small molecules bind to proteins. A powerful new field of science known as *structural biology* emerged in the twentieth century that enabled such visualization of the atomic-level detail of proteins. The field of structural biology was born on November 5, 1895, when William Roentgen made a profound discovery that would alter the course of medicine far into the future. Roentgen was the head of the physics department at the University of Würzburg in Germany, and he observed a type of electromagnetic radiation that had never been seen before. He attacked this mystery ferociously, and on December 28, 1895, he announced to the world his discovery of what he named *X-rays* to signify their mysterious nature.[1] Roentgen went on to win the first Nobel Prize in Physics for this discovery in 1901.

THE BIRTH OF X-RAY CRYSTALLOGRAPHY

Michael Pupin, a member of the faculty at Columbia University, confirmed Roentgen's findings, and the topic was aggressively taken up by physicists around the world.[2] Pupin's contributions to physics were such that the current physics building at Columbia was named Pupin Hall in his honor. I remember taking my first physics class on Saturday mornings in this building when I was a junior in high school, participating in the Columbia Science Honors Program. A skilled lecturer named Gene Dwyer illuminated the world of particle physics for me, which stimulated me to pursue a career in science, although my interests eventually shifted from particle physics to the interface between biology and chemistry.

The nature of X-rays was quite mysterious and remained so for some time. In 1912 Max von Laue proposed, and then demonstrated, that X-rays were another form of light; in other words, they were a type of electromagnetic radiation.[3] Specifically, he knew that passing light through an appropriate size filter would generate an interference pattern, in which the wave property of light would allow it to form a series of peaks and troughs in intensity, much like two waves meeting each other on the ocean. This interference is most notable when the spacing of the lines on the filter is about the same as the wavelength of the light used.

Laue speculated that X-rays could be a type of high energy light, with tiny wavelengths compared to the visible light emitted by a lightbulb. He realized that if this hypothesis were correct, it should be possible to observe an interference pattern by passing X-rays through a filter. However, the problem with this idea was that it was extremely difficult to test, because the wavelength of X-rays was predicted to be so small as to require a filter with spacing on the size of single atoms. It would be an impossible engineering problem to construct such a filter, given that an atom is about three hundred million times smaller than an inch.[4]

Laue proposed, however, that a crystal should serve as a perfect material to diffract X-rays and cause an interference pattern, because the spacing of atoms in the crystal was the same size as the predicted wavelength of X-ray light.[5] The subsequent demonstration of this phenomenon was convincing evidence that X-rays were in fact another form of electromagnetic radiation, with a small wavelength.

The British scientist William L. Bragg figured out how to use this phenomenon of X-ray diffraction by crystals to determine the atomic structure of the crystal. He demonstrated, along with his father William H. Bragg, that this concept could be applied to crystals of common table salt (sodium chloride, NaCl) as well as diamonds. In 1915 Bragg shared the Nobel Prize in Physics with his father. At the age of 25 the younger Bragg was, and remains, the youngest person to have ever won the Nobel Prize.[6]

THE APPLICATION OF X-RAY METHODS TO PROTEINS

One of Bragg's students, Desmond Bernal, became interested in applying Bragg's X-ray crystal analysis method to proteins.[7] This was a risky project, and not something most researchers would undertake. It was believed that proteins had no intrinsic structure, and there was little chance that they would diffract X-rays cleanly, in a specific pattern.[8] Thus the X-ray analysis of protein crystals was viewed as being an extreme long shot, in terms of scientific programs.

Crystals of inorganic materials had been known and studied for millennia; however, protein crystals represented a relatively recent discovery. In 1840 it had been found that protein crystals could be produced from blood, and subsequently these crystals, made of the abundant protein hemoglobin, were produced from many different species.[9] Hemoglobin is a protein that binds to oxygen and carries it from the lungs to other tissues in the body.

James Sumner showed in 1926 that he could purify a specific enzyme (urease) and cause it to form crystals. This allowed him to demonstrate using a variety of tests that enzymes are proteins. Sumner's method allowed the American biochemist John Howard Northrop to purify and crystallize other proteins, such as pepsin, which aids in digesting food in the stomach.

At the turn of the century, the chemical composition of proteins had been elucidated. It was known that proteins are assembled from repeating units of amino acids; the great German chemist Emil Fischer and the physiologist Franz Hofmeister demonstrated that these repeating units of amino acids in proteins are connected by a special linkage, which they termed the *peptide bond*.[10] The nature of this bond would become a crucial factor in the race to understand how these amino acid units fold up to make a structured protein. In a parallel approach, Fred Sanger would show that proteins had unique amino acid sequences—that is, each protein has a particular order of amino acids that is specific to the protein.[11]

In 1934 Bernal was able to obtain the first X-ray image of a protein, the protease enzyme pepsin, working with his student Dorothy Hodgkin. Hodgkin then established her own independent career, and published an image of the X-ray diffraction of crystals of the protein insulin, which controls blood sugar levels and is defective in some types of diabetes.[12] She went on to win a Nobel Prize in Chemistry in 1964 for her work on X-ray analysis of proteins.[13]

Shortly thereafter, another scientist, the young Max Perutz, moved from Austria to England, where he joined the laboratory of Bernal, who was known to his colleagues as "the Sage." Perutz began applying X-ray methods to the study of hemoglobin. Perutz later recalled his introduction to the field as follows: "I asked the Great Sage, 'How can I solve the secret of life?' He replied, 'The secret of life lies in the structure of proteins, and there is only one way of solving it, and that is by X-ray crystallography.' "[14] Within a short period of time Perutz was given a fair degree of independence, and he hired a young graduate student named John Kendrew.[15] They worked together on the structure of hemoglobin and myoglobin, using X-ray methods. Progress was slow, but Perutz loved the work, his new country, and his new colleagues and friends.

In May 1940, Perutz's research with Kendrew was interrupted. He was arrested on a clear Sunday morning by a police officer and taken to a small school, where he was locked up along with many other foreigners. After a week he and the other prisoners were moved to a cramped set of houses. From there, they were moved to a seaside complex on an island, where they were mass vaccinated over and over again with the same needle, potentially exposing each of them to any infectious agents carried by the others.

The prisoners were then loaded aboard the crowded troop carrier Ettrick, and transported to Quebec City, Canada. The Canadians stripped Perutz and the other prisoners naked and confiscated their personal items. Surely Perutz could only wonder what alternative universe he had crossed into. It eventually emerged that the British authorities, in response to public pressure, had decided that resident aliens were a danger during the war and had ordered foreigners like Perutz to be rounded up and imprisoned.

Eight months after being arrested, Perutz was finally returned to England. He later recalled his deep and lasting bitterness at this time of imprisonment: "To have been arrested, interned and deported by the English, whom I had regarded as my friends, made me more bitter than to have lost freedom itself."[16]

Nonetheless, Perutz's work flourished upon his return to his lab in England, and in the fall of 1947, with the support of Bragg, he established a research group to probe the structures of biological molecules using X-ray methods.[17] With the assistance of other skilled researchers, Perutz proceeded to develop the methods needed to elucidate the atom-by-atom structure of protein molecules. He was assisted in this effort by the physicist-turned-biologist Francis Crick. They were competing against the chemist Linus Pauling at Caltech,[18] who was pursuing his own program to determine the structures of proteins using X-ray methods.[19] Pauling was a well-known chemist who worked on atomic structure and had recently begun to explore the use of X-ray methods to analyze proteins.

Pauling was able to reason out one of the major stable structures that amino acid chains can form—the so-called alpha helix, beating Perutz in this race.[20] A key factor in Pauling's success was his chemistry background, which made him realize that the linkage between adjacent amino acids, the peptide bond, had to be planar, or flat, instead of bent.[21] Perutz and his team, all biologists and physicists, overlooked this fact, which caused them to miss the possibility of peptides forming an alpha helix.[22]

Perutz proceeded to verify Pauling's prediction using X-ray diffraction of protein crystals, confirming and extending this seminal discovery; he greatly regretted his failure of creativity to have suggested the existence of the alpha helix and his lack of chemical sophistication.[23] This is one of the subtle but critically important issues in multidisciplinary work. It is usually impossible to be an expert in all aspects of research problems that span multiple disciplines. The most successful researchers either manage to be staggeringly broad in their knowledge, such as Pauling was, or more commonly, to find collaborators in complementary disciplines who can augment their deficiencies. If Perutz had found such a chemist collaborator early on, he might have come up with the proposal for the alpha helix before Pauling.

Crick, working in Perutz's research unit, then applied X-ray methods to the analysis of DNA. Crick was then joined by the young James Watson, and together they tackled the problem of the structure of DNA, the genetic material. With data provided by Maurice Wilkins and, unknowingly, by Rosalind Franklin, they determined the double helical structure of a common form of DNA, this time beating Pauling's group.[24] This success at divining the secret of heredity brought worldwide recognition and funding to structural approaches, and to Perutz's specific research, which enabled his X-ray studies to expand.[25]

Perutz and Bragg continued to work together on the problem of protein structures. For while solving the structure of DNA was clearly important, the molecules that carry out the functions of a cell are proteins. Moreover, the structures of proteins are all different, and more complicated than DNA. At this point, Perutz teamed up with my future colleague Vernon Ingram and discovered a method that represented a vast improvement in their ability to use X-rays to determine protein structures. They found that by adding a heavy metal to the protein, they could solve a longstanding problem with the X-ray crystallography approach—the phase problem.

This problem refers to the fact that electromagnetic radiation, such as light or X-rays, has two aspects: the intensity and the phase.

You can picture these two aspects by imagining waves on the ocean. When two waves collide (which is similar to what happens when X-rays are diffracted by a crystal), each wave has an intensity (the height of the wave) and a phase (whether it collides at the peak or trough of the wave). The phase problem refers to the fact that only the intensity of the X-rays is collected by the detector in the X-ray crystal diffraction experiment, but the phase information is lost.

Solving this problem was essential for reassembling the atomic-resolution picture of a protein. Together, Ingram and Perutz showed that the heavy metal derivatization method (tagging the protein with a heavy metal) was a solution to this fundamental problem. Perutz later recalled this discovery as the most exciting moment of his scientific career.[26] In due time this breakthrough would allow Kendrew and him to determine the first detailed pictures of protein molecules.

Despite the conceptual importance of this breakthrough, the method of modifying a protein with different heavy metals was tedious. A much-improved solution to the phase problem was developed by Wayne Hendrickson, now a University Professor at Columbia, and a number of other researchers. This improved method used the highly intense X-rays emanating from national synchrotron facilities, originally built by physicists to study the deepest structures of atoms. The X-rays emerging from these particle accelerator facilities were initially an undesired waste product, but were harvested through the efforts of Hendrickson and others to yield a vast improvement in the ability of scientists to rapidly determine protein structures.

Perutz and Kendrew used their method to show that hemoglobin and myoglobin from different species have remarkably similar structures, providing another confirmation of Darwin's theory of evolution.[27] Ultimately it would be revealed that there is a striking similarity in the three-dimensional shapes of many proteins from nearly all species, owing to their shared evolutionary history.[28] Since proteins have to carry out similar functions in different species, it is not easy to change their shapes through mutation, without disrupting their function. This is why similar proteins from different species retain similar shapes.

In 1962 Watson, Crick, and Wilkins were awarded the Nobel Prize in Physiology or Medicine "for their discoveries concerning the molecular structure of nucleic acids and its significance for information transfer in living systems."[29] A few weeks later, Perutz and Kendrew won the Nobel Prize in Chemistry for their elucidation of the first protein structures using X-ray methods. Thus the Sage was proven correct, in that X-ray methods were crucial for revealing the fundamental nature of the molecules of life, as well as the basic mechanisms governing heredity and cellular physiology.

A MAGNETIC VIEW INTO PROTEIN STRUCTURE

A rival method was simultaneously being developed for probing protein mysteries, although it would not bear fruit until many years later. At the same time that Perutz and his colleagues were using X-ray crystallographic methods to elucidate the structures of proteins, a completely independent method for determining protein structures was being developed that involved the magnetic properties of atoms. This method was pioneered by I. I. Rabi.

Israel Isaac Rabi was born in what is now the southern part of Poland and moved to the United States when he was three years old; he settled on the Lower East Side of New York City, where he became fascinated with the local library.[30] He began methodically reading every science book in the library, in alphabetical order. For his bar mitzvah on his thirteenth birthday he entertained his family and friends with a formal lecture titled "How the Electric Light Works." He excelled in high school, earned a scholarship to Cornell University, and graduated with a bachelor's degree in chemistry.[31] He then worked for some time as a chemist in Manhattan but returned to Cornell to pursue graduate work in physical chemistry. Unfortunately, he did not enjoy his time at Cornell and he transferred to Columbia, back in New York City instead. At Columbia, Rabi switched his focus to physics, and finally found his niche for the first time in his life.[32]

Rabi developed both theory and experiments supporting the discovery of nuclear magnetic moments. Everyday magnets have a tendency to align with an external magnetic field; nuclei of atoms have a similar tendency. This propensity is called a *nuclear magnetic moment.* The nuclear magnetic quality of atoms can be measured experimentally.

In his first experimental confirmation of this idea Rabi measured the nuclear "spin," as it is known by physicists, of sodium.[33] Within a short time, he discovered the property of magnetic resonance, a process in which he could cause nuclei to move into an excited state by making use of their magnetic quality. This result would lay the foundation for the fields of nuclear magnetic resonance (NMR) spectroscopy, and magnetic resonance imaging (MRI). Rabi published this result in 1938, and went on to win the Nobel Prize in Physics in 1944. In 1964 he was the first faculty member in Columbia's history promoted to the exalted position of University Professor, which allows one to participate and teach in any department.

Just about the same time Rabi was doing his pioneering work, Felix Bloch, working at Stanford, independently discovered nuclear magnetic resonance.[34] Bloch began his training with the great physicist Neils Bohr, and subsequently trained with, or interacted with, many of the world's leading physicists, such as Wolfgang Pauli, Werner Heisenberg, Enrico Fermi, and Peter Debye.[35] He rose rapidly in stature in the physics community, and would have become a scientific leader in Germany; however, when Hitler was appointed Chancellor in 1932, Bloch, who was Jewish, fled Germany and moved to Stanford.[36] While there, he discovered this property of nuclear magnetism, which would in time become a powerful tool to elucidate the detailed atomic structure of proteins (see Figure 7.1). In 1952 Bloch and E. M. Purcell, who independently discovered nuclear magnetic resonance, were awarded the Nobel Prize in Physics for this transformative discovery.[37]

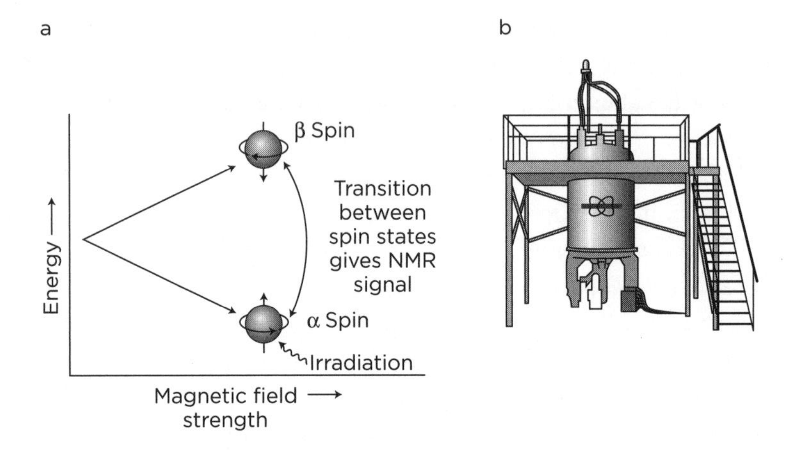

Figure 7.1 Nuclear Magnetic Resonance. (a) NMR spectroscopy involves determining the energy needed to excite the nuclear spin of atoms in a molecule. This can allow determination of the structure of the molecule. (b) NMR makes use of large magnets and sophisticated instruments for measuring these properties of molecules.

The use of NMR spectroscopy to illuminate protein structures emerged only slowly. In 1957 this new technology was used to measure the first spectrum of a protein. Martin Saunders, Arnold Wishnia, and John Kirkwood reported a 40 MHz NMR spectrum of ribonuclease, an enzyme involved in degrading RNA molecules.[38] They observed just four peaks in the spectrum, and tentatively concluded that these peaks might represent the different chemical environments that hydrogen atoms attached to the protein find themselves in. In other words, some parts of the protein are a little different from other parts, and these differences could be detected by NMR. The unique chemical environments of different hydrogen atoms reflect the fact that proteins adopt a stable three-dimensional shape, and each part of that shape can potentially have a unique environment.

These four peaks didn't provide much structural information, but they represented the beginning of a new approach to studying proteins. More advanced methods were developed to analyze protein NMR experiments, and complicated magnetic manipulations were designed to detect which amino acids were near each other in space.

Ultimately, it became possible to elucidate the complete three-dimensional structure of proteins (see Plate 1) using extremely sophisticated NMR methods. Kurt Wüthrich and Richard Ernst, both Swiss chemists, were separately awarded Nobel Prizes for their discovery of these high-powered methods that enabled the use of NMR for determining protein structures.

NMR is now routinely used to obtain high-resolution pictures of the three-dimensional shapes of proteins, albeit of small to moderate size proteins for the most part. One important difference between NMR and X-ray crystallography for protein structure elucidation is that X-ray methods study protein in the crystalline form, whereas NMR can study proteins in solution. Since proteins function in cells, and their normal environment is in solution, the solution information can be quite useful and complementary to the structural information obtained by X-ray crystallography.

Recent advances in solid-state NMR are enabling the exploitation of this method to challenging types of proteins, such as hydrophobic, membrane-localized proteins and protein aggregates that occur in neurodegenerative diseases, such as Alzheimer's disease, Huntington's disease, and Parkinson's disease. While both solid-state NMR and X-ray crystallography study proteins in a solid form (as opposed to in a solution form), solid-state NMR can study any form of a solid, not just a crystalline material. Crystalline materials are highly ordered, tightly packed regular arrays of molecules that are difficult to create; thus the greater flexibility of solid-state NMR implies that this method will be useful even in cases where protein crystals can't be created, but other, less organized, solid forms of proteins can be obtained.

After NMR and X-ray crystallography were developed for revealing protein structures, a vast array of proteins and protein complexes were studied with these methods, from the proteins making up the common cold to the proteins responsible for generating energy within each of our cells, in mitochondria. There are now more than 56,000 atomic resolution structures found in the worldwide protein data bank.[39] This is a freely accessible database and Web site that stores all of these protein structures and makes them available to scientists, as well as the general public. Indeed, these methods have become sufficiently robust and reliable, and several consortia have been formed to perform rapid protein structure determination. In other words, these consortia are producing, purifying, and solving the structures of hundreds of proteins in a semiautomated fashion using extensive robotic instrumentation.[40]

FROM STRUCTURES TO DRUGS

It was an article of faith among the early practitioners of this new field of structural biology that knowledge of the atom-by-atom picture of protein molecules would ultimately help to unravel disease mechanisms and to create new therapeutic agents. In the case of sickle-cell anemia, these prophetic statements would prove partially true for the first time, at least in terms of understanding disease.

Sickle-cell anemia was first described in detail in 1910 by Robert Herrick in Chicago.[41] He began his paper with the provocative statement "This case is reported because of the unusual blood findings, no duplicate of which I have ever seen described." He went on to describe a young African American man, age 20, with extensive ulceration of the legs, who had developed a bad cough, fever, and a chill. Upon examination of his blood, Herrick found that many of his red blood cells had a peculiar sickle shape. Indeed, Herrick was at a complete loss to explain the unusual appearance of these blood cells and how they might relate to the symptoms of this patient.[42]

In 1949 Linus Pauling showed that the hemoglobin protein from sickle-cell patients was different from the hemoglobin protein found in normal individuals.[43] To demonstrate this, Pauling measured the movement of the two protein preparations in an electric field, a type of analysis known as *electrophoresis*. Proteins migrate in an electric field in a manner that depends on their size, charge and shape. The observation that hemoglobin from normal individuals and those with sickle-cell anemia migrated differently demonstrated that these protein molecules must have some chemical difference between them, and it further suggested that this difference might be the cause of the disease. In fact, this method is still used to diagnose sickle-cell anemia.

In 1957 Vernon Ingram made his great discovery of the specific chemical difference between normal and sickle-cell hemoglobin. In his report he stated, "I have now found that out of nearly 300 amino acids in the two proteins, only one is different; one of the glutamic acid residues of normal haemoglobin is replaced by a valine residue in sickle-cell anemia haemoglobin."[44] This mutation causes the mutant protein to aggregate, and thereby causes the red blood cells to assume their sickle appearance, and to become more fragile.

Some years later, in 1976, Christopher Bedell and Peter Goodford at the Wellcome Laboratories in the United Kingdom followed up on Vernon's discovery by creating small molecules that were designed to fit into the sickle hemoglobin protein and improve its stability, with the hope of preventing sickle-cell disease.[45] Perutz built upon this promising start. At a meeting in Washington in 1980, discussions with colleagues led to the notion that he could try to co-crystallize with hemoglobin some small molecules that had been shown to prevent sickling of red blood cells. In this way it might be possible to understand the mechanism of action of these small molecules, and even to design more effective analogs that were similar in structure but that were better able to bind to the sickle hemoglobin protein.[46]

Perutz succeeded in obtaining the co-crystal structures of hemoglobin with several compounds, showing for the first time that it was possible to visualize in precise detail how a small molecule drug binds to a protein. Although this was exciting from a basic mechanism and technique standpoint, these compounds did not have all of the other properties needed to become drugs, so they ultimately did not prove to be effective drugs for sickle-cell anemia. For a small molecule to become a drug, it needs to be stable, nontoxic, properly distributed throughout the body after oral administration, and easy to manufacture, purify, and formulate. Most compounds, including these studied by Perutz, lack one or more of these attributes. Nonetheless, this was an important proof of concept experiment, and it represented the start of the field of structure-based drug design.[47]

Ultimately, the first commercial success at structure-based drug design was realized with captopril, an inhibitor of the angiotensin-converting enzyme (ACE), which is now used to treat hypertension, or elevated blood pressure.[48] This success began with the work of David Cushman and Miguel Ondetti at the Squibb Institute for Medical Research as early as 1968.[49] At the time, there was some debate about the physiological significance of the renin-angiotensin system in regulating blood pressure, since the protein renin was not consistently elevated in patients with high blood pressure, as would be expected for a hormone that regulated this process. In addition, the inhibitors of the renin-angiotensin system that were available were not effective and had modest, if any, effects on controlling hypertension.

Cushman and Ondetti, however, pursued this hypothesis, and were able to purify peptides from snake venom that were specific inhibitors of the angiotensin-converting enzyme.[50] Subsequently, they used their knowledge of the active site of this enzyme to design even more potent and selective, orally deliverable small molecule inhibitors of this protein.[51] This led to the eventual design, synthesis, and testing of Captopril, which created the market for ACE inhibitors for treating hypertension. As a class of drugs, inhibitors of this enzyme eventually generated more than $6 billion in annual sales and still benefit millions of patients with high blood pressure.[52]

Remarkably, Captopril was designed by making inferences about the structure of the ACE protein based on the X-ray structure of a different, but related, protein—carboxypeptidase A. This approach of designing a drug based on the structure of a similar protein, is challenging, but can be successful if the two proteins are similar enough in their three-dimensional structures. This process is analogous to transferring a computer program from one computer to another. If the two computers are running identical operating systems, the computer program will work flawlessly on both. If both are running slightly different versions of Windows, the program will probably work reasonably well. However, if one computer is running Windows and the other computer is running the Macintosh OS, it is unlikely that the computer program will work correctly when switched from one computer to the other.

Perutz's original vision of drug design and development using X-ray structures of a small molecule bound to its exact protein target was finally realized in the case of carbonic anhydrase II. By visualizing the atomic-level detail of the interaction of small molecules with this enzyme, it was possible to create dorzolamide, which is used for the treatment of glaucoma.[53] By inhibiting carbonic anhydrase II, it became possible to reduce fluid secretion into the eye, which reduces pressure in the eye, the problem in patients with glaucoma that causes damage to the eye.[54] Thus it finally became true that visualizing the structures of proteins made it possible to create new medicines that were engineered to interact with them. This has now become a commonly used method for discovering new drugs by most major pharmaceutical companies.

The ascendance of this structure-based drug discovery method is evident in the string of successes the method has produced, including the neuraminidase inhibitor Tamiflu (oseltamivir), used to treat flu infections, including the 2009 H1N1 flu pandemic, the renin inhibitor aliskiren for treating hypertension, the BCR-ABL inhibitor nilotinib, which is effective against drug-resistant chronic myelogenous leukemia, and a series of HIV protease inhibitors that are effective for treating HIV infection.[55] Indeed, it is becoming difficult to develop a new drug in the current era without using structure-based drug design.

DESIGNING DRUGS IN A COMPUTER

For those proteins that meet the requisite characteristics for structure-based drug design, there is a growing arsenal of tools. Once an atomic-resolution structure is produced by either X-ray crystallographic or NMR methods, the challenge is creating a small molecule that can bind tightly and specifically to the protein of interest.[56]

In the last decade an increasingly effective approach, called virtual screening, has been developed for addressing this challenge.[57] This approach uses powerful computers to predict whether a given small molecule can fit tightly into a pocket on a protein of interest. Since this computer prediction process can be rapid and has only moderate operating costs (although there are substantial upfront fixed costs), it is now possible to test in a computer the ability of millions of candidate small molecules to bind to a protein of interest. The virtual screening process was elegantly used in the industrial setting by Mark Murcko and colleagues at Vertex Pharmaceuticals, a successful pharmaceutical company founded by Josh Boger in 1989, based on the notion of structure-based drug design.[58] The key steps in virtual screening are (1) to build in a computer a collection of small molecules of interest, (2) to test these molecules in the computer for their ability to bind to a protein of interest, and (3) to test in the laboratory the best of the predicted compounds for their ability to bind or inhibit the protein of interest. Continued improvements are being made in each of these steps, further increasing the accuracy, speed, and utility of virtual screening.[59]

This virtual screening approach is being adopted by most pharmaceutical companies. It is a promising technology, although it is still limited by the fact that it works well for some proteins, but not at all for others, because predicting binding is extremely challenging to do precisely; unfortunately, the success rate for a given target protein is usually not something that can be predicted in advance.

In addition, the virtual screening approach is often limited by available computer processing power. An interesting solution to the problem was developed by W. Graham Richards, in which he and his collaborators made use of the idle processing power of 1.5 million personal computers. This project was modeled on the successful SETI@home precedent, in which home computers were used to sift through mountains of radio telescope data for signals of extraterrestrial intelligence. Richards and his colleagues were able to achieve this massively parallel computation by having volunteers download a specially designed screen saver to their computers and allowing it to run when their computer was not in use. Using this parallel computing approach, in 2002 he reported that they had been able to test, the ability of 3.5 billion different small molecules to interact with a dozen different proteins.[60] This was an effective demonstration of the power of distributed computing for solving complex computer processing problems.

BUILDING DRUGS ONE FRAGMENT AT A TIME

Yet even virtual screening cannot ever hope to assess the effectiveness of all possible small molecules against a protein of interest.[61] The truly vast immenseness of chemical space precludes making even a significant dent in the testing of all possible small molecules. A second recent innovation, fragment-based drug discovery, begins to address this challenge.

The rationale for fragment-based screening stems from the relationship between complexity and affinity of small molecules, articulated by Michael Hann and his colleagues in 2001.[62] Using a highly simplified model of a small molecule binding to a protein, they demonstrated that there is an optimal complexity for small molecule ligands. In this case, complexity is used to indicate the number of features on the compound, such as charged groups, hydrophobic patches, or other ways of interacting with proteins. As a small molecule becomes more complex, it is easy to see intuitively that the chance of it having a perfect fit in a pocket on a protein becomes vanishingly small. In other words, when molecules are too complex, the chance of finding any that actually bind to a particular pocket on a protein of interest becomes small.

On the other hand, when small molecules are extremely simple, with little complexity, it is easier to find some that match a particular protein binding pocket, but the affinity and specificity of the interaction is likely to be low. The affinity, or strength of the interaction, will be low because there are few interaction points between the small molecule and the protein (given that it is a simple compound of low complexity). The specificity will be low because it will be likely that such a simple compound can interact with other proteins as well.

Thus Hann's conclusion was that when looking for small molecules that can bind to a protein of interest, there is an optimal level of complexity. In fact, he suggested that traditional approaches have erred on the side of testing compounds with too much complexity. The solution in this case is to test simpler compounds. He quotes Albert Einstein in saying that "everything should be made as simple as possible, but not simpler."[63]

A few years prior to Hann's paper, a pioneering study was published in *Science* by Stephen Fesik and colleagues from Abbott Laboratories that showed how effective simple compounds could be. In this paper, published in 1996, they described a new technique, called SAR by NMR. SAR stands for structure-activity-relationship, and NMR is an abbreviation for nuclear magnetic resonance spectroscopy. Thus this technique was a way to define the relationship between chemical structure and binding affinity (activity) by using NMR analysis. The approach built upon their many years of experience studying small molecule interactions with proteins by nuclear magnetic resonance spectroscopy, NMR.[64]

The Fesik strategy was to initially test simple fragments for their ability to bind to a protein using NMR. These fragments are unusually simple compounds—essentially what you would get by fragmenting a typical drug into three or four pieces. They found two simple fragments that bound with low affinity; furthermore, when they stitched these two fragments together to make a more complex molecule of the typical size of drugs, they found the larger molecule bound with high affinity to their protein of interest. Using NMR, they were able to obtain an atomic-level picture of the interaction of this new compound with their protein of interest, which would aid in further modifying the compound.

This was an exciting study. It demonstrated that by starting with simple fragments of molecules, it was possible to iteratively build up a drug-like molecule that could bind with high affinity to a protein. Since there are far fewer possibilities for the number of possible fragments than fully assembled compounds, you can explore a much larger fraction of possible compounds in this way, potentially finding a solution to tackling undruggable proteins.

On this point, in 2009, Lorenz Blum and Jean-Louis Reymond reported a fascinating result—the construction of a computer database that contains all possible small molecules with 13 or fewer atoms of the elements carbon, nitrogen, oxygen, sulfur, or chlorine; this database contains 970 million small molecules.[65] While this sounds like a large number, it is vastly smaller than the number of possible larger drug-like compounds, which as noted earlier, is on the order of 10^{60}, or a thousand trillion trillion trillion trillion times as many compounds as the database contains. Moreover, since fragments are stitched together eventually to make larger compounds, you are effectively exploring a much larger number of molecules using fragments. In other words, if a typical drug is composed of three fragments, then by testing a thousand fragments, you are effectively exploring a billion drug-size compounds (1,000 × 1,000 × 1,000).

The Fesik paper launched the field of fragment-based screening, which has exploded in popularity and productivity in the last decade. This turns out to be a highly efficient method for discovering small molecules that can bind to proteins. The challenge in using fragment-based screening, virtual screening, structure-based drug discovery, or a combination of these methods, however, is that they are highly stringent: to be applicable, the target protein must be amenable to X-ray crystallography or NMR structure determination. This requirement stems from the need to figure out how to stitch the fragments together to make a full-size drug molecule; while structural information on the target protein isn't absolutely required, in practice such information vastly increases the chance of success. Moreover, these methods don't work for all proteins.

The percentage of proteins that can be successfully targeted even with these emerging methods is still unfortunately relatively small, although it is likely to increase over time. A major challenge is to determine whether these or other methods are ultimately applicable to the majority of the proteins that control physiological and disease processes. Is it possible to control the many ways in which these thousands of proteins talk to each other? Is it possible to control protein-protein interactions, which lie at the heart of living systems?

8

THE NATURE OF INTERACTIONS BETWEEN PROTEINS

Many proteins that control disease processes but are considered undruggable are not enzymes. This means that these proteins do not have an obvious pocket on their surface in which a small molecule can bind. Instead, these proteins function in the cell by interacting with other proteins. Blocking such protein-protein interactions has been considered to be one of the most challenging, if not outright impossible, tasks for drugs. In order to understand why, we need to explore the nature of protein-protein interactions.

THE QUEST FOR INHIBITORS OF PROTEIN-PROTEIN INTERACTIONS

One person who significantly aided in tackling the problem of protein-protein interactions was the biochemist James A. Wells. In 1973 Wells received his bachelor's degree in biochemistry from the University of California at Berkeley. From there he continued to build his expertise in biochemistry during his doctoral thesis at Washington State University and during his postdoctoral work. Armed with his training as a card-carrying biochemist, he made the somewhat unusual decision for a promising scientist at that time to pursue a career in the nascent biotechnology industry, at a young company named Genentech.

Genentech was an unusual biotechnology company—it was uniquely interested in, and successful at, promoting an academic culture in an industrial setting. Genentech, now part of Roche, is relatively young for a pharmaceutical company, but old for a biotechnology company, having been founded in 1976 by Robert Swanson, a venture capitalist, and Herbert Boyer, a biochemist. The company was founded on the then-new technology known as recombinant DNA. This technique allowed for the cutting and pasting of DNA molecules to create new gene sequences, which could then be used to synthesize entirely new proteins. In addition, it allowed for the large-scale production of proteins that were too scarce to be isolated from natural sources.

Shortly after the company's launch, in 1977, Genentech scientists were able to produce a human protein, somatostatin, using their new technology. The company then sold stock to the public in an initial public offering (IPO); the stock quickly skyrocketed from $35 per share to $88 within an hour. There was, perhaps justifiably in retrospect, enormous enthusiasm surrounding the promise of Genentech's approach.

In 1982 Genentech succeeded in bringing to market their first product, recombinant human insulin for treating diabetes, which was licensed to the large pharmaceutical company Eli Lilly. Human proteins are normally difficult to purify in useful quantities; by taking the human gene for insulin and putting it into bacteria, it became possible to make and purify large amounts of the human insulin protein from bacteria. It was in this year that Wells joined the company, seeing the promise of both the approach and the company. Wells rose through the ranks at Genentech, remaining there for 16 years. During that period, he built the protein engineering department, which focused on creating better versions of therapeutic proteins. Wells also worked out an approach to tackling some protein-protein interaction with small molecule inhibitors, discussed later in this chapter and in the next chapter.

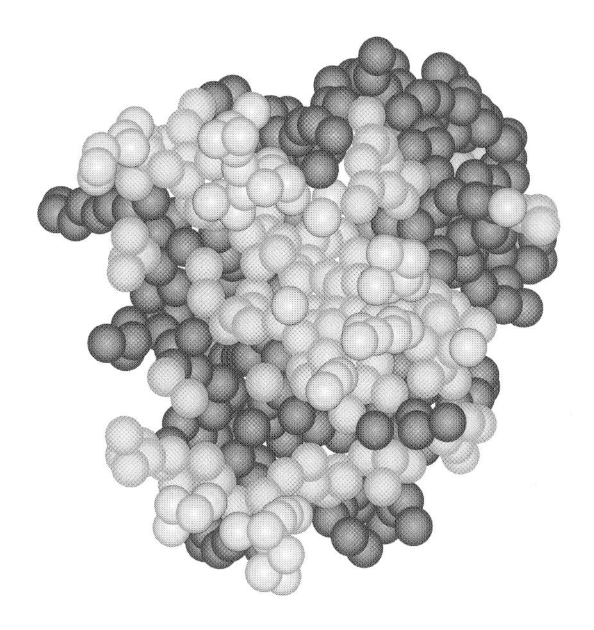

Plate 1 **Example of a high-resolution protein structure.** The three-dimensional structure of myoglobin is shown with greasy amino acids in yellow, charged amino acids in blue, and other amino acids in white.

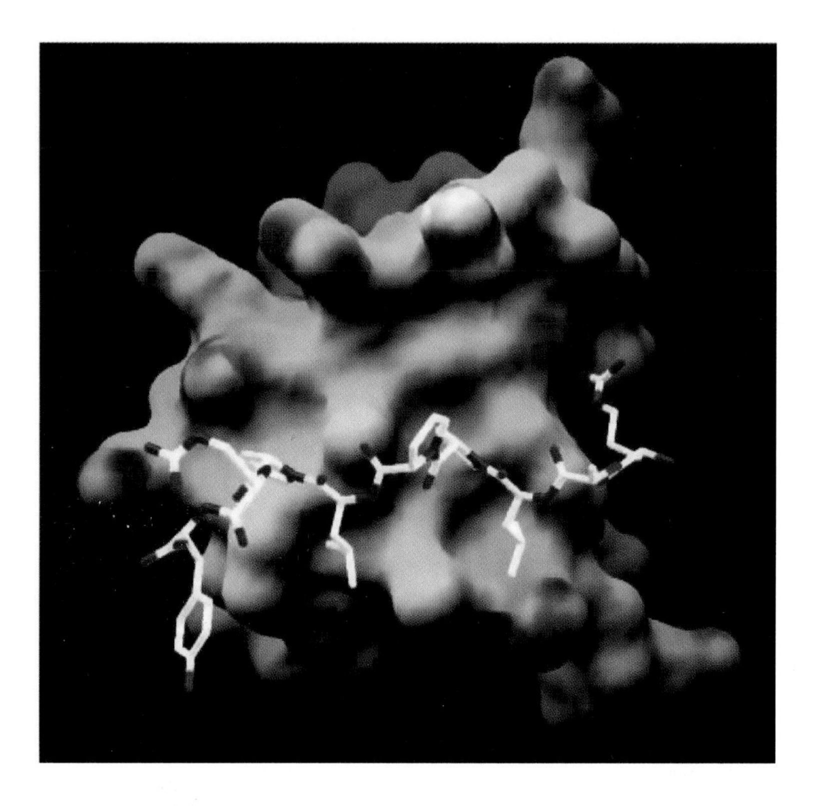

Plate 2 **Peptide interaction with a protein.** Shown is the interaction of a part of a protein known as the Src SH3 domain with the peptide it interacts with. *Source:* Mayer, B.J., SH3 domains: Complexity in moderation. *J Cell Sci* 114 (7), 1254 (2001). Reproduced with permission.

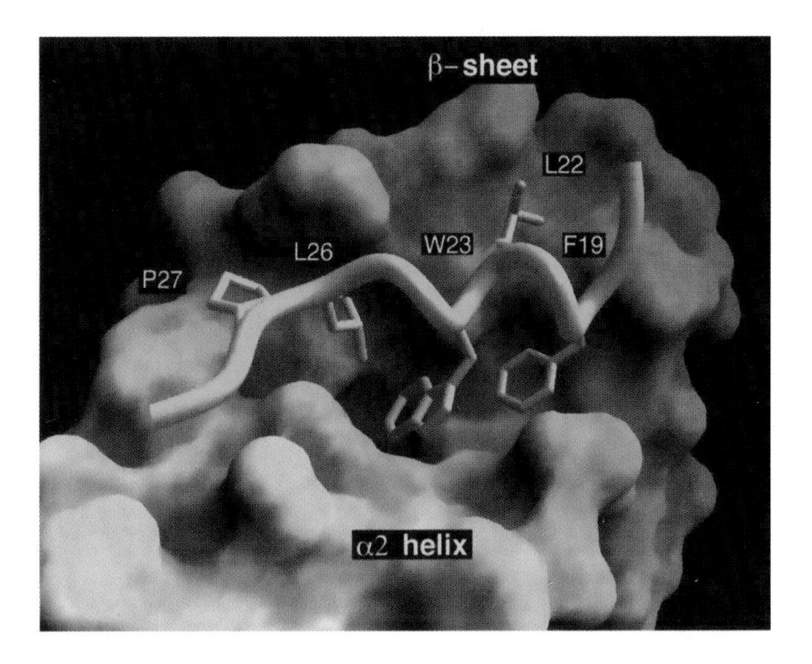

Plate 3 **The interaction of p53 with MDM2.** Shown is the peptide from p53 (in yellow) that interacts with MDM2. *Source*: From Kussie, P. H., Gorina, S., Marechal, V., Elenbaas, B., Moreau, J., Levine, A. J., and Pavletich, N. P., Structure of the MDM2 oncoprotein bound to the p53 tumor suppressor transactivation domain. *Science* 274 (5289), 951 (1996). Reprinted with permission from AAAS.

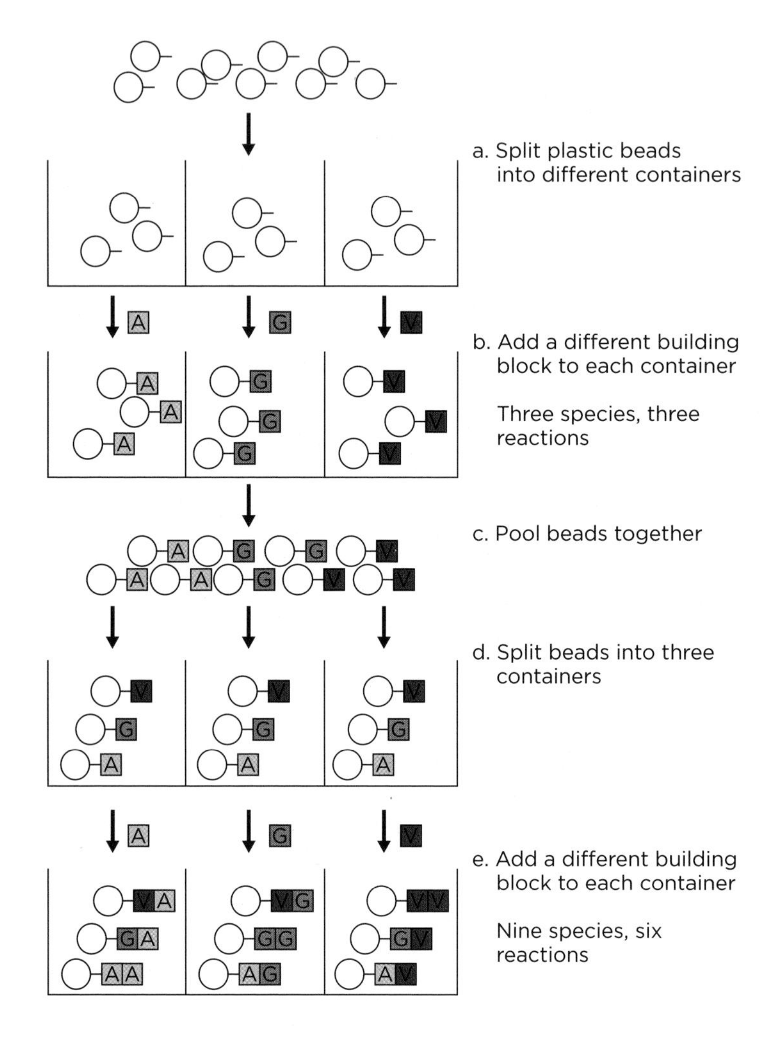

Plate 4 **Split-pool synthesis.** Split-pool synthesis involves splitting beads into different flasks, performing reactions in each flask, and then pooling the beads together again. By repeating this process several times, it is possible to make a large number of molecules with a small number of chemical reactions. *Source:* Stockwell, B. R., Frontiers in chemical genetics. *Trends Biotechnol* 18, 451 (2000). Reprinted with permission.

In 1998 Wells left Genentech to found a new start-up drug discovery company, Sunesis Pharmaceuticals, which was focused on the strategy of structure-based design of small molecule inhibitors of protein-protein interactions. Just a year later, Wells's pioneering efforts at drug discovery were recognized by his election to the National Academy of Sciences. Then in an unusual move, in 2005, after 23 years in industry, Wells made the move back to academia. He is now the Harry W. and Diana V. Hind Professor of Pharmaceutical Sciences, and Chair of the department at the University of California at San Francisco (UCSF). Wells is also the founder and director of the UCSF Small Molecule Discovery Center.

I first met Jim Wells in 2007 when I invited him to present a seminar on his research in my department. The visiting seminar is a tradition that most science departments have—notable researchers from other places around the world are invited to come present an hour-long seminar on their research accomplishments. The visiting speaker then spends the rest of the day meeting with other faculty members in the institution, hearing about their research programs. Usually there is an opportunity for students and postdoctoral scientists to have lunch with the speaker, since some of these students may want to do future research in the visiting scientist's lab—it is a time for such students to get to know the speaker and see if he or she would be a good mentor. Finally, the day ends with a casual dinner out with the speaker and several faculty members, discussing science, politics, culture, and whatever topics emerge as common interests. In the end there is a fairly small community of researchers working in any given research area, and over time they get to know each other well through these visits, and through many interactions at conferences.

Wells's most notable achievements have been in the area of blocking protein-protein interactions with small molecules, a journey described in the next two chapters. How important are these interactions? Nearly everything we do is regulated by protein-protein interactions, from the muscle contractions needed for walking, to the energy metabolism involved in digesting food, to the neuronal functions required for reading this book. The vast majority of processes inside cells also involve protein-protein interactions, including such activities as sending messages from one location in a cell to another, turning genes on and off, and committing cell suicide.[1] Protein-protein interactions are also involved in maintaining the structure of cells, in the transcription of DNA into RNA, and in the translation of RNA into protein. Moreover, a number of diseases involve specific defects in protein-protein interactions, such as Alzheimer's and other neurodegenerative diseases, many cancers, infectious diseases, and numerous other pathological processes.

THE NATURE OF INTERACTIONS BETWEEN PROTEINS

A protein-protein interaction occurs when one protein molecule physically contacts another protein molecule, in a specific arrangement.[2] Imagine an arm of one protein sliding into a narrow groove on a second protein, like a screw sliding into a hole. This is just one of the numerous ways in which proteins can interact with each other to yield stable or transient complexes.

While it is relatively straightforward, at least in principle, to target enzymes with small molecule drugs, protein-protein interactions are another matter entirely. This is unfortunate given the ubiquity of protein-protein interactions. If researchers could find a means of disrupting protein-protein interactions with drugs, it would open up an enormous number of possible targets for drug discovery; this could result in the discovery of new medicines for every disease imaginable.

Put another way, just about all of the 85% of proteins that are not currently considered druggable engage in protein-protein interactions. If we could find a way to tackle these interactions in a comprehensive, general way, we could solve the problem of creating drugs against nearly all proteins. In order to do this, researchers need to know how protein-protein interactions form, and how they might be disrupted with drugs.

So how do individual proteins assemble into larger protein complexes? They use high-affinity, strongly interacting protein-protein contacts. The physical basis of these interactions was suggested 70 years ago by Linus Pauling and Max Delbruck—they argued that protein-protein interactions must be caused by two proteins having complementary surfaces, in terms of the atoms found on those surfaces.[3] In other words, each protein presents a surface with complementary shape, charge, and greasiness to the other protein, allowing the surfaces to snap together into a tight-fitting complex.[4]

Thus the most straightforward view of protein-protein interactions is that two proteins have regions of their structures that are complementary, and when two such complementary regions encounter each other, they naturally assemble into a protein-protein complex. In this view, proteins are rigid bodies that associate without changing their shapes upon binding to each other. This is true in many cases, such as the interactions between antibodies and their target proteins, and those between the protease trypsin and its inhibitor, PTI (protein trypsin inhibitor).[5]

However, in some cases proteins change their shapes substantially upon interacting with other proteins. Proteins are not always statue-like, rigid bodies. Some proteins are flexible and dynamic and can assume different shapes depending on the environment. A good example of this is the interaction between the proteins trypsinogen (trip-SIN-oh-jin) and PTI. Trypsinogen is the precursor to trypsin, which is a protease involved in digestion. Trypsin, like other proteases, acts as an enzyme that breaks other protein molecules down into smaller pieces. Proteases function like molecular scissors; they have a propensity to cut protein molecules at specific amino acid positions. Trypsinogen, the precursor protein, is inactive and cannot cut proteins.

Trypsinogen contains a disordered, floppy region that adopts an ordered, rigid structure upon interacting with PTI. This same transition occurs when trypsinogen is activated to trypsin, which is the active form of the protease. Thus, in this case, trypsinogen has two possible shapes, or conformations, and binding to PTI induces trypsinogen to switch from one conformation to the other.[6] Protein-protein interactions can either involve the association of two rigid body proteins in a way that complementary surfaces meet each other and provide the strength of attachment, or they can involve one or both of the proteins adopting a different shape, which enables two complementary surfaces to interact, providing the glue for the protein-protein interaction.

THE CHALLENGE OF INHIBITING PROTEIN INTERACTIONS WITH SMALL MOLECULE DRUGS

One difference between drug-protein interactions and protein-protein interactions is the surface area of the binding interaction. Drug-protein interactions occur over a small area, usually 300 to 400 square angstroms.[7] An angstrom is a unit of length on a small scale, which is appropriate to the size of individual molecules. One angstrom is defined as 10^{-10} meters, or one ten-billionth of a meter (a yard is about the length of a meter). It was named after Anders Jonas Ångström, a pioneer in measuring the properties of molecules.

To get a feel for lengths at this scale, 1 angstrom is about the radius of an atom, 1.5 angstroms is the length of a carbon-carbon bond, 5 angstroms is the width of an alpha-helical segment on a protein, and 7 angstroms represents the length of a simple sugar molecule like glucose. When we measure surface area, we use units of square angstroms, just like measuring the size of a room in terms of square feet.

Protein-protein interactions occur over a larger area than the interactions between small molecules and proteins. Protein-protein interaction surfaces are in the range of 1,100 to 3,000 square angstroms, or 3 to 10 times larger than the surface area used in small molecule interactions with proteins.[8] In other words, when small molecules interact with proteins, they have tightly focused binding on a small part of the protein's surface. On the other hand, when two proteins interact with each other, they spread their binding interactions over a large surface. This makes it challenging to mimic the full extent of a protein-protein interaction with a small molecule and is one of the reasons why it is hard to make small molecules that block protein-protein interactions—it is hard to cover the full range of the protein-protein interaction with a small molecule and pick up enough binding energy.

Blocking protein-protein interactions could be especially important for controlling signaling mechanisms. In most cases signaling mechanisms involve the formation or disassembly of protein-protein interactions, which eventually leads to changes in gene expression, cell fate, or protein secretion. These changes in cellular processes are ultimately reflected in the functioning of an organism, and in the development of disease.

THE PROTEIN INTERACTIONS OF THE RAS PROTEINS

Returning to the example of the undruggable RAS proteins and their central role in many cancers, protein-protein interactions are crucial for both the normal signaling functions of RAS proteins and the cancer-causing functions of mutated RAS (see Figure 8.1). There is a protein that binds to RAS and turns it off. There is another protein that binds to RAS and turns it on. There are also a number of proteins that bind to RAS in its GTP-bound, "on" state and transmit signals to other parts of the cell. These include the proteins RAF and RALGDS (which stands for RAL guanine nucleotide dissociation stimulator), among others. Thus protein-protein interactions are fundamental and crucial in regulating RAS functions, and in mediating the cancer-causing effects of mutant RAS.[9]

To treat cancers with mutant RAS, or other cancer-causing proteins, we need to find a way to disrupt these protein-protein interactions. In order to do that we need to understand more about the detailed nature of these interactions. How do they work exactly and how can they be disrupted? To a large extent, protein-protein interactions can be classified into two major types: (1) interactions between two proteins that fold into stable structures on their own and

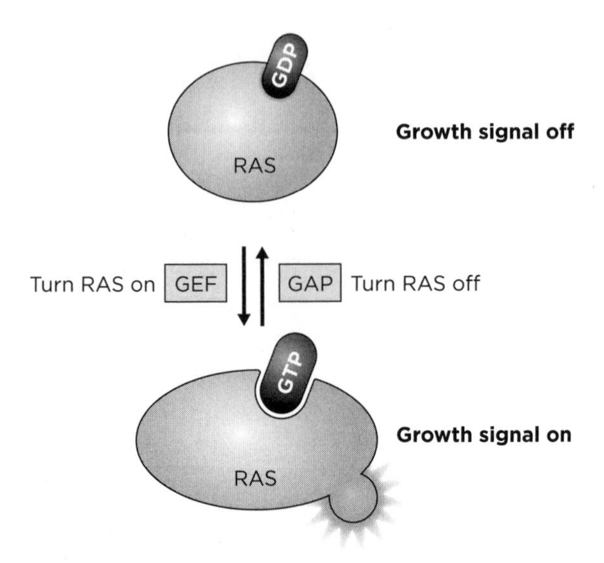

Figure 8.1 Detailed view of the RAS on-off cycle. RAS interacts with GEF (guanine nucleotide exchange factor) and GAP (GTPase-activating) proteins that activate and inactivate the growth-promoting function of RAS, respectively. Moreover, downstream signaling events also involve protein-protein interactions with RAS.

(2) interactions between a folded protein and a small, unstructured part of a protein.[10] The first type turns out to be the more challenging to block, but progress has been to some extent on both of these types of interactions.

One approach to disrupting the second, more tractable, type of protein-protein interaction involves using a fragment of one protein to break up the interaction between the two proteins (see Plate 2). For example, imagine a politician standing by the exit of the subway or train station, grabbing the hand of each person who comes outside and vigorously shaking their hand. If you wanted to prevent this politician from grabbing your hand, you could instead extend a mannequin's fake hand in place of your own, and then let go. The politician would grab the mannequin's hand in place of yours, and his hand would be fully occupied, so he would be unable to grab your hand, just as a protein that is bound to an artificial peptide would be unable to bind to its normal peptide partner.

Now to extend the analogy a bit, if you wanted to prevent the politician from grabbing anyone's hand, you could throw millions of mannequin hands into the air around the subway exit, so that the politician would only see the mannequin hands, and no actual hands of real people—there would be so few real hands compared to mannequin hands that the chance of the politician grabbing a real hand at random would become vanishingly small. In this case, the key to breaking up the interaction between the politician's hand and your own is creating a small artificial piece of your hand, which is the part of you that the politician is trying to grab onto.

The same approach can be applied to protein-protein interactions. When two proteins interact in a modular fashion, the part of one protein that interacts with the other can be produced separately and added to cells to compete for binding, preventing the interaction. This is a strategy for blocking the interaction between proteins and small, unstructured peptides. However, a major challenge with this approach is that small pieces of proteins don't generally fold correctly. To understand why this is so, it is necessary to understand something about protein folding.

Proteins are linear chains of amino acids that form what looks like a long string. As this linear chain is produced by translating a messenger RNA, the string of amino acids begins to fold, like a piece of string crumpling up into a complicated shape. Proteins generally have one low energy conformation, meaning a three-dimensional structure they will assume once they reach their most stable state. Generally speaking, a folded protein structure has the linear chain of the protein zigzagging this way and that, folded up onto itself in a complicated way.

Creating a small piece of a protein that might block a protein-protein interaction involves selecting a part of the protein sequence to make. Such a small piece of protein is called a peptide. Proteins typically are made up of 100 or more amino acids, whereas peptides are usually composed of between 2 and 20 amino acids. For example, you could create a peptide in the laboratory consisting of just the last five amino acids of a protein. If this last little bit of the protein was responsible for interacting with another protein, you could test whether that 5-amino-acid peptide would be enough to bind to another protein, and would disrupt the interaction between the two proteins.

Will such a peptide be functional? Will it act like the mannequin's hand in the example above, and break up a protein-protein interaction? First, the two proteins would have to interact in a way that mainly involved this five-amino-acid sequence on one protein interacting with the other protein. This means that the rest of the protein should not be involved, other than this five-amino-acid segment. In such a case, if you were to make this five-amino-acid peptide and then add it to cells, it should be able to replace the terminal region of the real protein in its interaction, and thereby disrupt the protein-protein interaction. This would realize the long-sought goal of disrupting protein-protein interactions, at least for this special case in which the interaction involves a peptide from one protein sticking into another protein.

However, there are three problems that prevent this strategy from being realized. First, peptides generally cannot permeate into cells—they are too polar to cross the greasy cell membrane that surrounds all cells. It would be like trying to have table salt pass through a layer of olive oil. It is quite difficult to get charged, polar molecules like peptides across the greasy cell membrane.

Second, even if there were a way to get the peptide into cells, it wouldn't necessarily have any structure. Proteins normally fold into complicated structures. Small peptides are unlikely to fold correctly, although in some rare cases they do. Peptides can usually adopt many possible shapes, and only a small number of these, maybe just one, is able to bind to the target protein. As the peptide continually changes shape, sampling all of these possible conformations, the chance that it will assume the one needed for binding becomes small. Therefore, even if the peptide could penetrate into cells, it wouldn't be able to bind to the target protein, because the peptide would be unstructured.

Third, peptides are susceptible to degradation by proteases. These molecular scissors are ubiquitous in cells and organisms, and when they come in contact with a peptide, they break it into pieces. Thus, for these reasons, it is not possible to disrupt protein-protein interactions with simple peptides.

A number of researchers have tried to overcome these problems with peptides by adding stabilizing elements.[11] Such constrained peptides are locked into a particular conformation—usually the conformation that is needed to bind to a target protein. This approach was used to try to tackle an important protein-protein interaction in cancer biology—the interaction between the proteins p53 and MDM2. A constrained peptide blocking this interaction was not suitable for becoming a drug, but it suggested that this previously intractable interaction in tumor cells might be amenable to disruption with a drug-like molecule. This turned out to be an important battleground in the long war on protein-protein interactions.

DISCOVERY OF THE P53 PROTEIN

The p53 protein was discovered by studying how the simian virus 40 (SV40) DNA tumor virus is able to transform cells. Unlike the RNA viruses studied by Harold Varmus and colleagues, DNA tumor viruses such as SV40 do not encode mutant versions of cellular proteins. Instead, such RNA viral proteins are specific to viruses—it was speculated that they acted in some way to affect normal cellular proteins.

SV40 was found to encode a specific viral protein, named the large T oncoprotein, which is able to assist in transforming normal cells into tumor cells. In 1979 David Lane and Lionel Crawford were working on this problem, and sought to determine which cellular proteins the SV40 large T oncoprotein could bind to—perhaps, they speculated, these cellular targets of large T could be proteins controlling the transformation of a normal cell into a tumor cell.

David Lane is, in many ways, a prototypical scientist—precisely because he didn't start out as one. He grew up in a large family and attended a strict Catholic elementary school, where he, in his words, "did not do very well, but rather scraped by."[12] He was told he was not "university material" by his teachers. He was rejected by five of the six universities he applied to. He gained admission to the sixth, University College London, where he studied biology as an undergraduate student from 1970 to 1973.

Lane recounted later how his father died of cancer at the relatively young age of 56, which must have had a profound effect on Lane's future research direction. Indeed, another cancer pioneer, Harold Varmus, specifically remarked in his autobiography that his mother's death from breast cancer inspired him to go into cancer research, not in a futile effort to cure the disease, but as a way to understand this mysterious biological process known as cancer.[13]

Lane was impressed with one of his professors, Avrion Mitchison, who displayed a great enthusiasm for, and knowledge of, broad areas of biology; Lane began his doctoral work under Mitchison's supervision in 1973, but his research was initially unsuccessful. However, with assistance and training from a postdoc in the Mitchison lab, Don Silver, Lane began to thrive. Fortuitously, around this time, Lionel Crawford, who was working at the Imperial Cancer Research Fund (ICRF), heard about a promising young PhD student named David Lane. Crawford had trained in biochemistry at Emmanuel College in Cambridge, England, and then had done postdoctoral work in virology at University of California at Berkeley and Caltech, before becoming head of the Department of Molecular Virology at the ICRF.[14]

Crawford had begun studying the mysterious large T oncoprotein of SV40, and realized that Lane's expertise would be invaluable for the project, because purifying proteins associated with large T would require the precise, new techniques that Lane had mastered. Lane enthusiastically joined Crawford for his postdoctoral work, thereby completing his PhD in less than three years. Within a few months, he was promoted again, being appointed as an independent lecturer at Imperial College, an impressive feat for a 25-year-old who was not "university material."

In 1979 Lane and Crawford reported in the journal *Nature* that the large T protein from SV40 formed a complex with a cellular protein that has an apparent size of 53,000 daltons.[15] A dalton is a unit of mass that is about equal to the mass of one hydrogen atom; thus, the protein found by Lane and Crawford was about 53,000 times larger than a hydrogen atom. The size of proteins is measured in these units of molecular mass and can be estimated by causing the protein to migrate through a gel material in the presence of an electric field. Since the gel is fairly dense, small proteins move quickly through the gel, but larger proteins get caught on the gel material and move more slowly. Thus, with appropriate standard proteins present, you can estimate the size of a protein by the speed with which it migrates through a gel. A small molecule drug is typically in the range of 200 to 600 daltons, whereas a peptide is in the range of 500 to 3,000 daltons. The smallest proteins, in contrast, are 10,000 or more daltons, and the largest proteins can be hundreds of thousands of daltons or even upwards of a million daltons. So this apparent 53,000 dalton protein that was associated with large T was a middle-sized protein, fairly average and unremarkable upon first appearance.

This new protein was reported to interact with viral proteins by others as well, including Daniel Linzer and Arnold Levine at Princeton, and teams led by Pierre May, Robert Carroll, and Alan Smith.[16] David Baltimore's lab and others showed that this protein, dubbed p53 because of its apparent molecular mass, was abundant in tumor cells, but not in normal cells. Suddenly, this unremarkable new protein began to appear quite interesting—it seemed to function as a tumor-causing protein, given that it was present in tumor cells, but not in normal cells. A mutant p53 gene was isolated from tumor cells and, when introduced into normal cells, found to transform normal cells into tumor cells. Thus a consensus emerged that p53 was a cancer-causing oncogene.

However, a series of experiments caused this conclusion to be questioned. David Wolf and Varda Rotter, and then other groups, reported that the p53 gene was defective in mouse leukemia cells. If the p53 gene helped to cause tumor formation, why would the gene be disrupted in tumors? It should be turned on and highly expressed, as had been seen in some tumor cells.

Another key observation was that one sample of the p53 gene, when introduced into cells, did not have transforming properties. This nontransforming clone had a different DNA sequence compared to transforming samples of p53—it turned out that the nontransforming sample was the normal, unaltered DNA sequence, and the transforming samples were all mutants of p53 that inactivated the normal p53 protein that was still present in cells.

Bert Vogelstein showed that human colon cancers have mutations or deletions in the p53 gene, again consistent with the idea that under normal circumstances, p53 prevents tumors from forming, and that the gene needs to be inactivated to allow tumor formation to occur. Finally, it was found that mice lacking p53 develop cancer, and that patients that have inherited mutations in p53 develop the disease Li-Fraumeni syndrome, which results in a greatly increased probability of developing cancer at a young age. Thus a consensus emerged that p53 is a tumor suppressor—it prevents normal cells from becoming tumor cells, and is inactivated by mutation or deletion in many tumors.

P53 blocks cell division and cell growth and can induce cell death; cancer is associated with a lack of cell death and an increase in cell growth. Therefore, you would imagine that the amount of p53 is tightly regulated in normal cells, to prevent these cells from accidentally being killed by p53. Indeed, a regulator of p53 turns out to be an important protein controlling tumor formation.

THE MDM2 PROTEIN

In 1991 a new protein was discovered as being greatly amplified and highly expressed in a mouse tumor cell line with tiny chromosomal pieces called double minutes (as in tiny, mai-NOOT, i.e., not the measure of time). The finding that this protein was highly expressed in a tumor cell line suggested it was oncogenic; indeed, later studies found that the gene encoding this protein was present in multiple copies in a large percentage of sarcomas, as well as other tumor types. This new protein was named MDM2, for murine double minute 2; the next year this new protein was reported by Arnold Levine and Donna George and their colleagues to interact with and inhibit p53, identifying the key to MDM2's ability to promote tumor formation.[17]

Within a few years, Arnold Levine was able to work with Nicola Pavletich at the Memorial Sloan Kettering Cancer Center to obtain an atomic resolution picture of the interaction between MDM2 and p53, using X-ray crystallography (see Plate 3). Such high-resolution pictures of a protein interaction are helpful in assessing how an inhibitor might be designed. Pavletich had studied chemistry as an undergraduate student at Caltech. He then was diverted towards molecular biology at Johns Hopkins University School of Medicine, receiving his PhD in molecular biology and genetics. He is now an accomplished structural biologist, with an ability to determine X-ray structures of challenging proteins.

Together in 1996, Levine and Pavletich, and their colleagues, reported in the journal *Science* that MDM2 bound specifically to a peptide region of p53 composed of just 15 amino acids. In other words, this turned out to be one of the more tractable types of protein-protein interactions in which a small peptide segment of one protein interacts with a second protein; this suggested it might be possible to make a drug that blocks this interaction.

Pavletich and his colleagues described the region of MDM2 that interacts with p53 as being a "twisted trough, having a cleft lined with hydrophobic amino acids."[18] The peptide from p53 forms an alpha helix, a kind of corkscrew shape, and slides into this trough on MDM2, inserting three greasy amino acids down into the hydrophobic bottom of the trough. When these hydrophobic surfaces contact each other, the interaction is stabilized and the two proteins stick together.

When you view a protein structure like this, and you see three hydrophobic amino acids (on p53) forming a tight interaction with hydrophobic regions on the target protein (MDM2), it is likely that this explains the binding interaction. However, it is important to test this notion with additional evidence.

One piece of evidence that these three hydrophobic amino acids on p53 are critical for binding to MDM2 is that these amino acids are exactly the same in the p53 molecules found in different species. That is, if you compare the amino acid sequences of p53 molecules from different species, they are different in some positions, but these three amino acids are always exactly the same. This is just what we would expect if those three amino acids played critical roles in the interaction with MDM2. If these amino acids are necessary for the stability of the MDM2-p53 interaction, then changing or removing these amino acids should prevent p53 and MDM2 from interacting. If these three amino acids are the same in all species, it implies that evolution has eliminated individuals that arose when these amino acids were altered, likely because completely disrupting the MDM2-p53 interaction would be lethal to the organism. It should be noted that deleting the gene for MDM2 in mice causes early death of mice before they are born, because p53 becomes massively overactivated. However, partially reducing the amount of MDM2 in mice, and thereby partially increasing the amount of p53, causes mice to be resistant to tumor formation. Thus there is a threshold level at which blocking MDM2 and activating p53 is beneficial, but completely inhibiting MDM2 with a drug would likely be toxic.

In terms of the size of the p53-MDM2 interface, it is only about 1,500 square angstroms, which is on the smaller side of protein-protein interactions. This was the first hint that it might be possible to find a means of disrupting this interaction between MDM2 and p53, thereby unleashing the tumor-destroying activity p53. If this were possible, it would open up a new therapeutic avenue in the 50% of human tumors that retain functional p53. Not only would p53-MDM2-interaction blocking drugs have potential for killing tumor cells on their own; they could also potentiate existing cancer therapies, many of which act through p53. As David Lane and Peter Hall concluded in a commentary in 1997, "Such agents would have enormous potential to modulate the response to conventional chemotherapeutic drugs and radiation."[19] Moreover, blocking the p53-MDM2 interaction with drugs would represent an assault on the protein-protein interaction problem, and would raise the hope that the grand challenge of tackling these undruggable protein surfaces might ultimately be met. How might these protein-protein interactions ultimately be targeted by drugs?

9

FROM PROTEIN-PROTEIN INTERACTIONS TO PERSONALIZED MEDICINES

FROM PERIWINKLE TO PROTEIN BLOCKER

For many years, the possibility of targeting protein-protein interactions with drugs seemed fanciful at best, and most researchers viewed such a goal as being impossible. However, a glimpse of a future in which it might be possible to control protein-protein interactions could have been seen by astute observers in 1957. That was the year that James Bertram Collip isolated an intriguing new molecule from the Madagascar Periwinkle plant in his lab at the University of Western Ontario, Canada.[1] This plant grows three feet tall and is adorned by white to pink flowers containing a bright red circular center.

Collip was a young biochemist who quickly rose to the head of his Biochemistry department. On a sabbatical in 1921 he visited the University of Toronto and made critical contributions in the successful discovery and purification of insulin, a major breakthrough that enabled clinical use of natural, purified insulin, before the advent of Genentech's recombinant protein.

It was many years later that a female patient directed Collip to the Madagascar periwinkle plant because of its common use in her native Jamaica to make a tea that was rumored to be effective in treating diabetes.[2] Collip and his associate Robert Noble began studying the plant to see if it might yield useful compounds. They found no compounds with a beneficial effect on blood sugar levels or diabetes, but they did find that extracts of the periwinkle plant induced destruction of bone marrow and reduced the number of white blood cells. They wondered if there could be a natural product in the extract that might be useful for treating leukemias.

Working with the bioorganic chemist Charles Beer, they succeeded in isolating an elaborate, complex natural product that they named vinblastine. Shortly thereafter, the chemist Gordon Svoboda at Eli Lilly was able to isolate a similar natural product that came to be called vincristine. Before long, the first cancer patients were treated with these new molecules, and their tumors responded favorably.

Success was short lived, however, as the patients' tumors grew back. Nonetheless, these were powerful molecules. Vincristine and vinblastine were developed into standard treatments for a variety of cancers, despite their side effects, which can include hair loss, nausea, constipation, and hearing loss, among others. Their mechanism of action was found to be unlike other emerging chemotherapeutic drugs of the time, in that these vinca alkaloids, as they came to be known, affected microtubules.[3] Microtubules are assemblies of tubulin proteins that form long chains and extend throughout the interior of cells. Cells make use of microtubules as a railroad network to move proteins and organelles around the cell. Motor proteins, functioning as small, automated trains, can glide down microtubule networks and transit from one location in a cell to another.

Vinblastine and vincristine prevent the correct assembly of tubulin proteins into the microtubule network in cells.[4] These drugs have a complex effect on tubulin and microtubules, by preventing the normal dynamic assembly and disassembly of tubulin molecules and breaking up tubulin polymers into isolated tubulin molecules. Thus, these natural products are able to disrupt the protein-protein interaction occurring between two tubulin molecules; this was the first hint that protein-protein interactions were not the impregnable fortress that many had assumed them to be.

Nonetheless, almost all protein-protein interactions remained impervious to small molecule inhibition for many years. The large size of protein-protein interfaces seemed daunting to researchers interested in finding small molecule disrupters of these interactions.

HOT SPOTS DISCOVERED

In the mid-1990s James Wells and his associate, Tim Clackson, discovered an unexpected feature of most protein-protein interactions that began to change this perspective. Clackson and Wells published a paper in *Science* in 1995 showing that the interaction between human growth hormone and its receptor was focused on a "hot spot."[5] This is the less tractable type of protein-protein interaction, involving two large protein surfaces contacting each other, as opposed to a short peptide from one protein interacting with another protein. However, Clackson and Wells discovered that while two proteins often contact each other through a large surface area, the entire interface does not contribute equally to the strength of the interaction between the two proteins. This had been a crucial implicit assumption. Just because two proteins contact each other across a large area, it doesn't necessarily follow that the entire area is needed for the strength of the protein-protein interaction.

Imagine taking a piece of 8.5-×-11-inch paper and gluing it to the wall with a single dab of glue in the center of the page. All of the paper is in contact with the wall. However, which part of the interaction is important for the strength of the interaction? Only the small area that holds the glue.

Clackson and Wells demonstrated that this principle of a hot spot applied to the interaction of human growth hormone and its receptor. These two proteins have 30 amino acid positions that interact on each side of the interface, spanning 1,300 square angstroms. Clackson and Wells tested the relevance of each of these 30 interaction points by individually mutating them, and thereby removing one interaction at a time. When they made these mutations, effectively testing the importance of each interaction alone, they found that most of the interactions had little effect on the strength of the interaction between the two proteins. Indeed, more than three-quarters of the binding energy was caused by a small patch of greasy amino acids at the center of the interaction, just like the dab of glue holding that hypothetical paper to the wall. Based on the disproportionate importance of this region, they termed it a hot spot for the interaction affinity.

A number of researchers have found hot spots in other protein-protein interactions.[6] Indeed, a survey of 22 protein-protein interactions by Andrew Bogan and Kurt Thorn at the University of California at San Francisco confirmed that hot spots are ubiquitous in these interactions.[7] The implication for small molecule drug discovery was that it might be possible to target hot spots with drugs to improve the chance of disrupting protein-protein interactions with small molecules. This was a crucial discovery of Wells, and it heralded a weak point in the impenetrable armor that surrounded protein-protein interactions. For the first time, there was a mechanistic rationale for targeting these interactions, and a hope that success might ultimately be possible.

Indeed, this approach was used to disrupt the interaction between interleukin-2 (IL-2) and its receptor (IL-2R) with a small molecule.[8] IL-2 is a secreted protein that controls the growth and survival of T-cells, a type of white blood cell that is part of the immune system. IL-2 functions by binding to its receptor, IL-2R, on the surface of T-cells, and initiating a signaling cascade inside these cells. The protein-protein interaction between IL-2 and its receptor was known to be crucial in the function of the immune system because disrupting this interaction with proteins has been shown to be effective in preventing the rejection of transplanted organs by the recipient patient's immune system; this led to approval by the U.S. Food and Drug Administration of antibodies that block this interaction, and their widespread clinical use. Antibodies are large proteins produced by the immune system with a potent ability to bind to almost anything. Small molecule drugs disrupting the IL-2 interaction with the IL-2 receptor would be preferable to antibodies, because they could be given orally in the form of a pill, instead of the repeated injections needed for antibodies.

In 1997 Jefferson Tilley and colleagues at the Roche Research Center of Hoffmann–La Roche in Nutley, New Jersey, reported that they had discovered a small molecule that could block the interaction between IL-2 and IL-2R, known as a cytokine interaction with its cytokine receptor. As they noted in their paper, published in the *Journal of the American Chemical Society*, "To our knowledge, this represents the first well-characterized example of a small molecule, non-peptide inhibitor of a cytokine/cytokine receptor interaction."[9] Indeed, this study showed that it is possible to design and engineer small molecules that can block what was originally viewed as an interaction impervious to such inhibition—between a secreted protein and a cell surface receptor. Subsequently, a limited number of other protein-protein interactions were successfully targeted with small molecules, suggesting that this disruption was not an isolated case.

A PERSONALIZED CANCER DRUG

In 1994 Lyubomir T. Vassilev and his colleagues at the Roche Research Center reported the successful discovery of a small molecule inhibitor of that crucial interaction in cancer biology, the interaction between p53 and MDM2. Their approach was to test a diverse collection of small molecules for their ability to disrupt this key interaction. Before long, they identified a novel compound with the ability to block the MDM2-p53 interaction—they named this new compound "nutlin" for Nutley inhibitor. This greasy molecule was found to bind to MDM2 in a manner similar to that of p53 by occupying the deep hydrophobic groove inside the region of MDM2 that interacts with a linear peptide sequence on p53, in the interfacial hot spot. Vassilev and colleagues found that nutlin could displace p53 from MDM2 and unleash p53, allowing it to carry out its tumoricidal activity, in both cells and mouse cancer models.

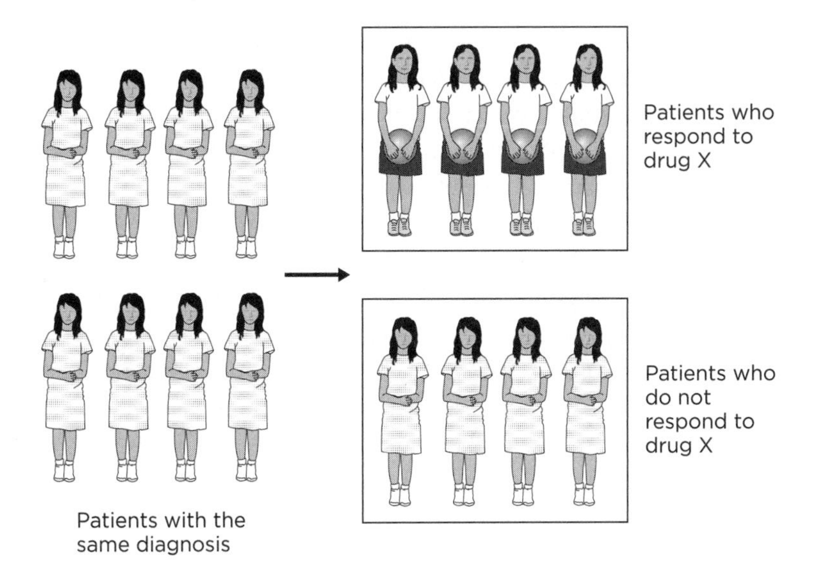

Figure 9.1 Personalized medicine. Patients with the same diagnosis can now be separated into different subpopulations based on genetic information. One approach is to use this genetic information to predict which patients will respond to a drug and which will not respond to the drug. This segregation prevents the non-responders from being exposed to the side effects of the drug unnecessarily.

As expected from its mechanism of action, nutlin is only lethal to tumor cells that maintain the normal, unaltered wild-type p53 protein. About half of all tumors have wild-type 53—they develop other mutations that allow them to be converted from normal cells into tumors. In such cells, nutlin disrupts the interaction between p53 and MDM2, allowing the free p53 that forms to accumulate and activate a cell death program. In contrast, in tumor cells in which p53 has been inactivated by mutation, disrupting the interaction between p53 and MDM2 has no beneficial effect, because the p53 in these cells is not functional.

Therefore, nutlin has the potential to become a type of personalized medicine—designed to be specifically effective in patients with a particular genetic background (see Figure 9.1). In this case nutlin would be particularly effective in patients who have tumors in which the p53 protein has not been mutated or otherwise inactivated. This has important consequences, because it suggests that optimally designed clinical studies with nutlin would involve measuring the p53 mutation status in patients' tumors before treatment begins. Only those patients likely to benefit from the drug treatment would be exposed to the potential side effects.

There is an additional complication, however. Some tumors that have wild-type p53 nonetheless have a mutation in another protein that impairs the response of the cells to p53. In other words, for p53 to activate cell death in tumor cells, it must be able to initiate a particular sequence of events in these cells. If other proteins involved in this chain of events are damaged by mutation, then p53 is unable to activate cell death. A p53-activating drug would not be effective in patients with these mutations.

These proteins function like a bucket brigade, in which a bucket of water is passed from person to person down a line of people connecting a fire hydrant to a fire. P53 is like the first person in the brigade, who fills up his bucket and passes it to the next person. If anyone in the bucket brigade is unable to pass the bucket along, the brigade breaks down and no water makes it to the fire. In the same way, if proteins downstream of p53 are inactivated by mutation in tumor cells, then p53 is unable to transmit its cell death signal.

Vassilev and colleagues found that this issue was potentially important in predicting which tumors might respond to nutlin treatment. For example, they noted that although few melanoma tumors have mutations in p53 (meaning that unaltered p53 is present in these cells), melanoma cells tend to have an impaired p53 response. This was determined by measuring how sensitive these cells were to radiation or DNA-damaging chemicals—these treatments usually cause cell death by activating p53. The fact that many melanoma cells are resistant to these p53-activating treatments suggests that somewhere downstream of p53 they have a mutation that blocks the p53 bucket brigade.

These researchers found that tumor cells containing a higher than normal amount of MDM2 were particularly sensitive to treatment with nutlin.[10] This could be rationalized by suggesting that these tumors had only used one mechanism for blocking p53—the increased abundance of MDM2, which binds and degrades p53. Therefore, in such cells, breaking up the interaction between MDM2 and p53 releases p53, so it can carry out its bucket brigade action, which results in tumor cell death.

The important implication of this finding was that there was an even better prediction about which patients might respond to nutlin treatment. Not only those patients with wild-type p53, but particularly the subset of patients also having unusually high concentrations of MDM2 in their tumors would be most responsive to treatment with nutlin. This defined a targeted group of patients likely to show the best clinical response to a nutlin drug.

This notion of personalized cancer therapy is increasingly transforming the landscape of cancer treatment. For example, women with early stage breast cancer have often been treated with radiation and surgery, but up to 30% of these patients have a later recurrence of breast cancer. If, in contrast, these patients with early-stage breast cancer are treated with more aggressive therapy, such as conventional chemotherapy, they are less likely to relapse. Because there is no way to predict which patients are the ones who will relapse, almost all of these patients are treated with aggressive chemotherapy, causing serious side effects and reduction in quality of life. If there were a way to stratify patients based on their risk of relapse, many could be spared unnecessary treatment.[11] This is an example of how personalized medicines could transform cancer treatment.

Personalized medicines have the potential to improve the patient experience, by reducing side effects and increasing the effectiveness of therapy. However, there is an anxiety related to this topic for some analysts in the drug discovery industry. This anxiety stems from the fact that as the suitable patient population for a drug is sliced thinner and thinner, the potential market for the drug, and the potential return on investment, becomes smaller and smaller.

We can see this principle at work in the nutlin inhibitors of the MDM2-p53 interaction. Not knowing which patients might respond, you would be tempted to give the drug to all cancer patients, in an extreme example, which is a potential $72 billion worldwide market. On the other hand, if you know that only patients with normal, non-mutated p53 will respond to the drug, then you reduce the market size to $36 billion. If you further target only those tumors with both normal p53 and overabundant MDM2, the drug would have a $3 billion to $4 billion market. While this is still substantial and would justify the investment to create this drug, in other cases the market might be so small that a company's management or investors might decide this would not be a reasonable investment of resources.

If additional genetic markers were found that could further sub-stratify patients, the probability of successful treatment would increase, but the market size would continue to decrease. In the extreme limit, each patient is unique, but it is not practical to develop a drug that can only be used by a single person, as the costs would not be justified by what the individual would pay for the drug. This is likely to become a significant driver of the cost of prescription drugs, as the pharmaceutical industry seeks to make shrinking markets as lucrative as possible to warrant continued development of personalized medicines.

In the case of the MDM2-p53 interaction, it has been possible to find a small molecule that can disrupt this protein-protein interaction, and this may lead to an effective personalized medicine. However, MDM2 interacts with proteins other than p53. Nutlin won't block all of the oncogenic activities of MDM2. To understand these other oncogenic roles of MDM2, it is necessary to understand how MDM2 functions biochemically.

THE DISCOVERY OF LIGASE PROTEINS

MDM2 is a type of enzyme known as a ligase. This means it is responsible for linking together two different molecules. Specifically, MDM2 causes a protein named *ubiquitin* to be linked to other proteins.

Ubiquitin is a small protein that is, as its name suggests, ubiquitous in cells. The discovery of this protein dates back to 1969, when the Israeli scientist Avram Hershko began his work as a postdoctoral fellow in the laboratory of Gordon Tomkins at the University of California in San Francisco, studying how the protein tyrosine aminotransferase (TAT) is degraded inside cells. Curiously, Hershko and others found that the degradation of TAT could be prevented by depleting cellular ATP energy stores. This was inconsistent with the prevailing paradigm, which was that protein degradation was a nonspecific process of decay, and not regulated in a precise way. Hershko reasoned that if ATP (cellular energy) was needed for degrading some proteins, then those proteins might be actively destroyed through a precise biochemical mechanism.

Hershko returned to Israel to establish his own laboratory, where he was joined by a graduate student named Aaron Ciechanover. Together they demonstrated that in extracts made from cells, they could detect protein degradation that was dependent on an unusually small protein with a molecular weight of just 9,000 daltons. This protein came to be called ubiquitin, because of its ubiquity, being present in virtually all species. Hershko and Ciechanover reported they were "astonished to find, however, that ubiquitin was bound by a covalent amide linkage" to these other proteins, which was unheard of at that time.[12] An amide linkage is a way of irreversibly connecting two proteins, essentially making them into a single protein.

Hershko, Ciechanover, and Irwin Rose found that linkage to ubiquitin was a signal that in most cases caused a protein to be degraded. Later, it was found that the linkage of a protein to ubiquitin can also act as a signal for other events, separate from causing protein degradation. This ubiquitin conjugation was revealed to be a major means of modifying proteins and controlling their fate.

The system that installs this linkage of ubiquitin to proteins is elaborate. There is a large class of proteins, known as ligase proteins, which are responsible for the addition of ubiquitin onto other proteins. There are predicted to be more than 600 ligase proteins in humans, which is equivalent to the number of kinases and proteases, both of which are large and important protein families.

MDM2 is one of these ligase proteins, and is responsible for attaching ubiquitin onto p53, inducing the destruction of p53. This is one of the ways that MDM2 blocks the tumor-suppressing activity of p53 in cells. Thus MDM2 causes p53 to be destroyed.

Nutlin acts by preventing the association between p53 and MDM2, thereby preventing MDM2 from conjugating ubiquitin to p53, which therefore prevents the destruction of p53. However, MDM2 also degrades other proteins; that is, MDM2 also tags other proteins with ubiquitin. MDM2 has been shown to have oncogenic, cancer-causing effects that depend on these other targets in addition to p53. Thus an ideal drug would target not just the association between MDM2 and p53, but would directly prevent the ability of MDM2 to tag any protein with ubiquitin. This would likely shut down all of the cancer-causing effects of MDM2, rather than just the effect on p53.

In 2005 a group lead by Allan Weissman and Karen Vousden reported a biochemical, test-tube-based screen for such compounds. This was a challenging project, because the activity of MDM2 depends on protein-protein interactions. Previously, the belief was that MDM2's ligase activity was not something that could be blocked with small molecules. That is, it was believed that the ligase activity of MDM2 was undruggable. However, by testing a large number of compounds, Weissman and Vousden were able to identify a few that could modestly block the ability of MDM2 to transfer ubiquitin to other proteins, including p53. These active compounds were not entirely specific for MDM2, and the concentrations needed were higher than was desirable, but they nonetheless demonstrated that it was in principle possible to find small molecules that might disrupt the protein-protein interactions needed for MDM2's ligase activity. The precise biophysical mechanism has not been elucidated for these compounds, and it remains a question of great interest to understand exactly what aspect of the mechanism is druggable.

Around this time, in my laboratory, we became interested in the possibility of blocking MDM2's ligase activity with small molecules.

I had become fascinated with this question of which proteins are druggable and which are truly undruggable. The huge class of ligase proteins was known to be crucial in biology and potentially valuable in medicine, since there are hundreds of these proteins involved in all aspects of cell physiology. Yet because they have been considered undruggable, there was little effort or success at creating specific small molecule inhibitors of ligases. Indeed, if ligases could be shown to be druggable, they would represent almost 20% of the druggable proteins in humans.

A graduate student in my lab, Ariel Herman, was able to develop an assay to rapidly test hundreds of thousands of compounds for their ability to inhibit MDM2 ligase activity inside cells. Testing the compounds in cells seemed particularly relevant to us, because all of the needed cofactors and accessory proteins would be present in suitable form. In contrast, when researchers normally use test tube biochemical assays with purified proteins, they have to decide exactly which proteins and cofactors to include in the assay, and in what form, creating an artificial situation that might not mimic what occurs in cells.

After testing a large number of compounds (~270,000), we found two that could block MDM2 ligase activity. This result confirmed for us that MDM2 ligase activity is not something that is easy to inhibit with small molecules; indeed, this is one of the more difficult screening assays, in terms of the chance of finding active compounds. Nonetheless, we were excited to find these two compounds that could inhibit MDM2 ligase activity. We are still evaluating their suitability as drug candidates, and the precise biophysical mechanism by which they act. Importantly, we now know that MDM2's ligase activity can indeed be inhibited if we use the right molecules and the right assay system.

In summary, it turns out that some protein-protein interactions are indeed druggable. Although it is more difficult to block protein-protein interactions using small molecules compared to the success rate with other proteins, the goal can be accomplished, at least in some cases. Is this an approach that can be expanded, systematized, and applied across the human genome? Is it possible to inhibit any of the 600+ ligases encoded in the human genome? At the limit, are the 85% of putatively undruggable proteins ultimately amenable to inhibition with small molecules? I think the answer is perhaps. Perhaps with the right molecules, we can unlock this vast reservoir of potential drug targets. The central question is: Which molecules are the right molecules?

10

A REVOLUTION IN PEPTIDE SYNTHESIS

A possible solution to finding the kinds of molecules that can block large numbers of protein-protein interactions, the most undruggable of targets, originated in 1949 in the laboratory of Dilworth Wayne Woolley. In that year a young biochemist, Robert Bruce Merrifield, joined the Woolley laboratory. Merrifield grew up in California. He received his undergraduate education at the University of California at Los Angeles (UCLA) and then worked as a laboratory technician, maintaining rat and chicken colonies, before starting graduate school at UCLA, where he met his future wife.[1] The day after he received his PhD he was married. The next day he and his new wife drove to New York so he could begin working immediately in the Woolley laboratory for his postdoctoral research.

Woolley was widely known for his pioneering work on vitamin biochemistry.[2] Moreover, Woolley had become diabetic early in his childhood, and subsequently lost his vision. He succeeded in science, nonetheless; his perseverance was aided in part by his wife reading scientific papers to him, an essential part of any scientist's work.[3]

THE CHALLENGE OF PEPTIDE CHEMISTRY

The new postdoctoral scientist Merrifield was assigned by Woolley to the task of purifying a small peptide that had growth-promoting activity. The chemical identity of this peptide was unknown, and there was interest in determining the composition of the peptide. Knowing the chemical identity of this peptide, it might become possible to make large quantities for medicinal purposes, as well as chemically similar peptides that could have superior properties. Finally, it might even be possible to figure out how this peptide was able to exert its growth-promoting activity in cells and to learn something new about the mechanisms governing cell growth. Thus figuring out the chemical makeup of this peptide was a key first step in understanding and improving upon it.[4]

Merrifield succeeded in isolating this peptide and determining that it was composed of five amino acids linked together, forming a pentapeptide. He determined the identity and specific order of these five amino acids: serine (SEER-een), histidine (HIS-te-deen), leucine (LOO-seen), valine (VAIL-een), and glutamate (GLOO-tu-mait). By comparing the five-amino-acid sequence of this pentapeptide to the complete sequence of the protein insulin, which had been reported by Fred Sanger in 1951, Merrifield found that his peptide was chemically identical to a small piece of the insulin protein.

Having successfully identified the chemical makeup of this peptide by purifying it from natural sources, Merrifield set about trying to create a synthetic version in the laboratory using chemical synthesis. This was a challenge, because methods for creating peptides were limited. Only a small number of active peptide sequences were known, and it was challenging to make them in the laboratory.

The first peptide derivative had been synthesized as early as 1882 by Theodor Curtius. The German chemist Emil Fischer undertook a systematic effort to show that these natural peptide molecules could be made in the laboratory. He made his first peptide—composed of just two amino acids—in 1901, prior to winning the Nobel Prize in Chemistry for his earlier work creating sugars and pieces of DNA. Slowly, other scientists built upon his painstaking success.[5]

It was in this climate that Merrifield undertook his synthetic chemistry challenge. Each natural amino acid had to be obtained or synthesized, and then each had to be joined together in a suitable order to create the correct peptide sequence. For each of these joining reactions, some chemical groups had to be protected so they would not create unwanted products during the reaction. Other chemical groups had to be activated so they would be capable of reacting to form the desired peptide bond. Finally, at each step the desired product had to be purified from side products, all of which represented a challenging and time-consuming process.

Eventually, Merrifield succeeded in synthesizing this small peptide, but it required a huge investment of his time and resources. For some people, experiencing a technical challenge like this turns them off to science, because of its difficult and painstaking nature. This response is certainly understandable, as it is difficult to spend years working on extremely challenging projects, especially when they are not successful. However, some scientists become energized by such a challenge and seek a way to overcome it. Merrifield was the latter sort, and he began to contemplate the woeful situation in this new world of peptide synthesis that he had entered.

Merrifield's peptide was made up of just five amino acids but it required a monumental effort to synthesize, consuming 11 months of work and with an overall yield of 11%—meaning that for every pound of material he started with, he ended up with less than two ounces of purified peptide. He knew that an octapeptide, composed of eight amino acids, had been recently synthesized by Vincent du Vigneaud's group, and it required the effort of six people over an extended period of many months. Merrifield was interested in the seemingly impossible—making a large number of peptides to optimize their activities. However, such a research program to find a more effective peptide was beyond his reach.

A NEW TECHNOLOGY IS BORN

In a stroke of brilliance that would have far-reaching effects—transforming both the drug industry and the field of synthetic chemistry—Merrifield conceived of a solution to the problem of synthesizing peptides.[6] His idea was to perform the synthesis of peptides on an insoluble material, such as a small plastic bead (see Figure 10.1).

In this radical approach the first amino acid would be attached to an insoluble plastic material. Only the amino acid would attach to the plastic bead; any by-products or other chemicals present in the reaction solution would not be attached to the plastic bead and would simply be washed away. In other words, Merrifield imagined he could pass the reaction with this insoluble bead through a small filter that would capture the bead but allow the rest of the reaction solution to be washed away. Then he would be left with the only part of the reaction he was interested in—the product, which would be attached to the bead.

Merrifield realized he could then attach a second amino acid onto the first one on the bead, and so forth, building a peptide on this plastic bead. At each step it would be feasible to separate the product of the reaction from by-products and other undesired materials. He imagined he would be able to create a complete peptide in short order, avoiding the painstaking purification work that would normally be needed. All of this could be accomplished by tethering the growing peptide onto an insoluble material, such as a plastic bead.

One way to think about Merrifield's invention is in terms of preparing an egg to eat. When you crack open a raw egg, sometimes a piece of eggshell gets mixed in with the liquid egg. Anyone who has tried to remove such a piece of eggshell knows it can be a difficult process, especially when there are many tiny pieces of eggshell. Imagine millions of extremely tiny pieces of eggshell mixed in with the liquid egg, and you will see how difficult it is to separate the egg from the eggshell if they have been mixed together in the liquid phase.

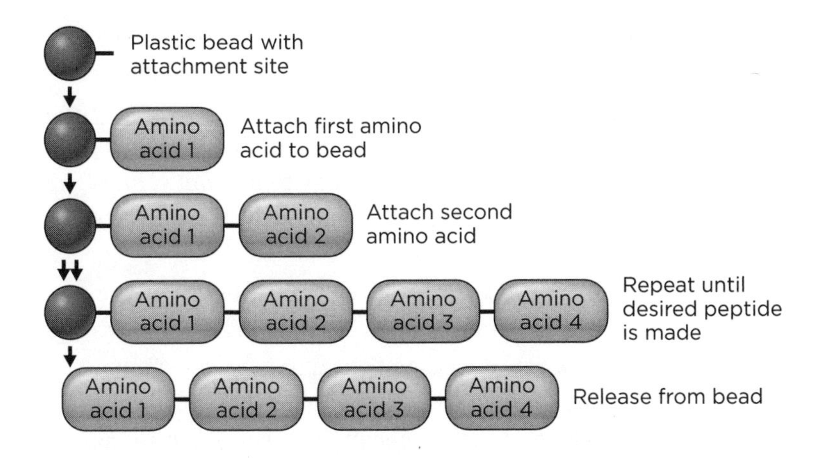

Figure 10.1 Solid-phase peptide synthesis. Solid-phase synthesis involves: (1) attaching an amino acid onto a solid-phase resin, or plastic bead; (2) attaching the next amino acid; and (3) repeating this process several times to create a complete peptide. In the last step (4), the molecule is released from the resin.

On the other hand, separating the eggshell from a solid, hard-boiled egg is not usually problematic. The hard-boiled egg is like the plastic bead that Merrifield used. The egg can be washed under the faucet and the tiny pieces of eggshell simply washed away, leaving behind the hard-boiled egg. It's not a precise analogy, because the next step would be to attach another piece of food to the hard-boiled egg, but it describes the general flavor of Merrifield's approach.

Merrifield began work on realizing this new approach to constructing peptides. To develop a completely new technology like this is a challenge. Funding agencies assume that a radical new idea like this is unlikely to be successful (probably because most radical new ideas are not successful), and it is therefore difficult to get funding for such high-risk research. Merrifield was fortunate to have been supported by the visionary Rockefeller University and allowed to pursue this new technology.

Eventually, Merrifield was able to make progress, and at a scientific meeting in 1962, he described his success at synthesizing a peptide made of four amino acids, using his new "solid-phase" method.

The peptide had been literally constructed on a plastic material without the need to isolate, purify, and characterize each intermediate in the traditional manner. Merrifield made steady progress from this point forward, creating a series of peptides rapidly and in good purity and yield. He pushed the technology to its limits and began synthesizing larger and larger peptides, and finally full-size proteins such as insulin.

Merrifield and his lab members were then able to create machines that could be programmed to carry out these reactions and automate the entire process of synthesizing peptides. Today, you can order long peptides for a few dollars from commercial suppliers that use such machines. Merrifield's ability to create an entire enzyme in the laboratory using this new methodology was stunning to the scientific community, and word of this breakthrough technology spread to the popular press.[7]

The traditional peptide synthesis community, however, did not rush to embrace Merrifield's new technology. At the time of Merrifield's original 1962 breakthrough, Joseph Fruton, a leader of this community, remarked, "This is not the way to synthesize peptides."[8] Another traditional peptide chemist said solid-phase synthesis "should be suppressed by the community."[9] Other peptide chemists were even more hostile in their comments. Part of this toxic reaction might have been because Merrifield was a biochemist, not a classically trained chemist, and there was skepticism from chemists that a biochemist could make important molecules such as peptides reliably.

I recall a similar cultural division that persisted in the world of chemistry even as recently as the 1990s, when I was a graduate student. Many traditional synthetic chemistry students (and perhaps some of the faculty) were skeptical that anything of value could emerge from biological chemistry studies. There was a sense of territorialism from traditional chemists, in which biological chemists and biochemists were not seen as genuine chemists. Fortunately, much of this division seems to have finally been erased in the current era, as the great synergies possible at the chemistry–biology interface have enriched the work of both synthetic chemists and biological chemists.

With all of these challenges, and much skepticism surrounding his approach, Merrifield persevered. Ultimately, in 1984, Merrifield's pioneering work on solid-phase synthesis was recognized with the Nobel Prize in Chemistry. Even after winning the prize, Merrifield was low key and accessible to those working in his lab. He thought of his lab members as family, and they viewed him with affection. He gave independent research projects to everyone in the lab, including the research assistants, who in other labs might be assigned more menial work. One research assistant, Anita Bach Rieman, began working for Merrifield right out of college and continued for 12 years. Her parents did not live in the area, and she later reminisced that she considered Merrifield her second father.[10]

Another research assistant, William B. Macaulav, recounted his own unusual experience with Merrifield.[11] Macaulav graduated from Trinity College and was seeking employment in a research laboratory in New York City. One day he entered the secluded campus of Rockefeller University, and inquired as to whether there might be any positions available. Macaulav was given a brochure describing the research labs at Rockefeller, and his interest was piqued upon seeing Nobel Laureate Robert Merrifield listed, as he had studied Merrifield's work in his biochemistry class in college. Upon expressing interest, he was surprised to find that he was offered an interview the same day.

Macaulav rushed to the library to learn every aspect of peptide synthesis before his interview. His ultimate interview with Merrifield centered, however, on sports, family, and hobbies; Merrifield wanted to determine if each lab member would fit seamlessly into the lab culture. This was more important than scientific talent, in terms of its impact on the lab. Merrifield hired Macaulav, and he succeeded in the lab. Macaulav is now the Anne Youle Stein Professor of Clinical Orthopaedic Surgery and Director of the Center for Hip Replacement at Columbia University Medical Center.

On May 14, 2006, Robert Bruce Merrifield passed away at the age of 84. Yet the technology he created continues to transform biology and medicine.

THE ORIGIN OF RESINS

Merrifield's revolution was built upon early pioneering work exploring the utility of solid-phase resins in organic synthesis. As the French philosopher Bernard of Chartres noted in 1159 (which was paraphrased by Isaac Newton in 1676), great scientists are like dwarfs on the shoulders of giants, who see farther than other men because they are carried high enough to prevent their vision from being obscured. In other words, all great scientific breakthroughs build upon prior less revolutionary work by other scientists.

For hundreds of years, scientists had been interested in natural resins; these are viscous secretions produced by plants that are insoluble in water. The light brown–hued material known as amber is a type of solid resin that is made into jewelry. Frankincense and myrrh are resins that have been used for thousands of years, going back to the ancient Greeks and Romans; they are also mentioned in the Bible. The ancient Egyptians made use of sandarac and mastic, resins that were made into paints.[12]

A major leap forward in the use of resins came with the invention of synthetic resin materials. In 1908 the French chemist Leon Grognot was awarded a U.S. patent (No. 906,219) "for the manufacture of resinous products capable of replacing natural resins."[13] He built a device and a process for combining formaldehyde with phenol to make a hard, brilliant material that can be "cast into thin sheets, beads or sticks, as desired," in the words of Grognot.

Chemically, phenol is a ring with an alcohol group appended to it. Formaldehyde is a small reactive molecule. When a molecule of phenol combines with a molecule of formaldehyde, it creates a slightly larger molecule with an alcohol group attached to it. In other words, the formaldehyde attaches with the phenol, causing the molecule to grow a little larger. This process happens over and over, and eventually a large super-sized molecule is created. Since the molecule is large, it is a solid rather than a liquid or gas. Generally speaking, the larger a molecule is, the more likely it will form a solid under normal conditions.

At the same time that Grognot filed his patent, the Belgian chemist Leo Baekeland was developing a similar process to make synthetic resins. He invented a formula that became popular, producing the first widely used synthetic resin, which he called Bakelite. This resin came to be used in radios, telephones, jewelry, cameras, machine guns, and toys, among other items. This new field of synthetic resins began to grow rapidly. In 1927 James McIntosh of the Diamond State Fibre Company in Bridgeport, Pennsylvania produced a new synthetic resin by combining two different chemicals—glycerol and phenol. McIntosh termed this new synthetic resin Acrólite.[14] Variations followed over the years, and synthetic resins have now been used to make many products, including laboratory countertops, circuit boards, and numerous coatings.

Synthetic resins were used in many different applications, but one was particularly critical for Merrifield's future work. In 1935 Basil Albert Adams and Eric Leighton Holmes produced synthetic resins that had striking effects on the softening of water. Similar to Grognot and Baekeland, they produced a polymeric, insoluble material that could be used for exchanging the ions magnesium and calcium in "hard water" with sodium ions. This exchange prevents calcium and magnesium deposits from forming in the water. Since sodium is much more water soluble, it stays in solution and the problem of these hard water deposits is alleviated. These "ion exchange resins" as they came to be called, were widely used in the water softening business. They also found uses in the processing of sugar from cane and beets, and ultimately in protein purification in biochemistry labs like Woolley's.

An important conceptual leap forward came in November 1948, when Alexander Galat reported that he could convert nicotinonitrile (NIH-koh-tin-oh-NIH-trayl) into the vitamin nicotinamide using a synthetic resin. The first inkling of the power of resins in synthesizing molecules can be glimpsed by reading Galat's terse account in the *Journal of the American Chemical Society*.[15] He recalled his thought process, remarking, "It appeared to us that by the use of a water-insoluble catalyst of basic nature it should be possible to avoid or to minimize the hydrolytic action, *while maintaining the catalytic effect* due to hydroxyl ions present on the surface of the catalyst."[16] This was a key intellectual step forward, because rather than using resins as physical materials such as coatings, Galat used resins as catalysts to assist in making molecules. This adaptation to synthetic chemistry would come to full fruition in the work of Merrifield 15 years later.[17]

SOLID-PHASE SYNTHESIS OF DIVERSE MOLECULES

After Merrifield revolutionized the field of peptide synthesis, researchers began to apply his solid-phase methods to the synthesis of other kinds of molecules, including oligonucleotides [oh-lig-oh-NOOK-lee-oh-tides] and carbohydrates. It is now possible to synthesize a large oligonucleotide in a few hours, using the kind of automated solid-phase chemistry that was originally used for peptide synthesis.

One of the early efforts to adapt solid-phase synthesis to other uses was that of Clifford Leznoff. In 1967 he joined the faculty of York University in Toronto, where he stayed until his retirement in 2005. As early as 1972 he reported on the use of polymeric resins in the synthesis of molecules other than peptides.[18] The challenge of making sugars using automated, solid-phase chemistry was solved due to the efforts of Peter Seeberger, a Professor at the Swiss Federal Institute of Technology (known as ETH) in Zurich, Switzerland.[19]

Nonetheless, despite the seminal quality of Merrifield's work and these subsequent efforts, they solved only half of the original problem, which was how to efficiently make a large number of peptides. The invention of solid-phase synthesis allowed Merrifield to make individual peptides more efficiently, but what about the challenge of making a large number of peptides at once? Solid-phase synthesis laid the foundation for solving this problem, but didn't directly address it. The solution wouldn't emerge until the early 1980s, when several researchers simultaneously attacked this problem.[20]

THE BIRTH OF COMBINATORIAL CHEMISTRY

In 1983 Ronald Frank, working in Germany, reported the use of cellulose paper as a solid material to enable the simultaneous synthesis of numerous oligonucleotides.[21] At the same time Richard Houghton, working at the Scripps Clinic and Research Foundation in La Jolla, California, developed a related technology, in which he enmeshed plastic beads into tea bags that could be sorted into different flasks, depending on which amino acid was to be coupled to the beads.[22] Similar to Frank's method, it enables the rapid synthesis of peptides.

In 1984 Mario Geysen, working in Australia, described a procedure for making hundreds of peptides using solid-phase synthesis.[23] Geysen created small plastic rods that served as solid supports for synthesis. The growing peptides were attached to these rods, which were arranged in a regular array of 96, corresponding to a commonly available miniaturized set of test tubes, with 96 vessels. By immersing each rod in a corresponding reaction vessel with the suitable amino acid to be coupled, it was simple and efficient to make a large number of peptides on these rods in a parallel fashion. This came to be known as parallel synthesis. For the first time, the solid-phase method was used not just to simplify the process of making one peptide, but also to enable the efficient synthesis of hundreds of peptides at the same time, in a parallel fashion.

Using automated parallel-synthesis, it became possible to make hundreds or even thousands of peptides. But what about making millions of peptides? Such a possibility was beyond even the reach of parallel synthesis.

In 1982 Árpád Furka conceived of a different and potentially more powerful approach to making peptide collections. Furka had trained as a postdoctoral fellow at the University of Alberta in Canada and returned to his homeland of Hungary in 1965, contemplating his recent research determining the amino acid sequences of protein enzymes. As Furka later related, he speculated about the huge possible number of proteins that could exist of a given protein size.[24] He realized that as a peptide or protein chain grows longer, the number of possible sequences becomes astronomical. To see this on a small scale, consider that there are 20 naturally occurring amino acids. Dipeptides are the smallest possible peptide, consisting of just two amino acids linked together. Since each of the two amino acids in the dipeptide could be any of the 20 naturally occurring amino acids, there are 20 × 20 = 400 possible dipeptides.

Using this type of calculation one can determine that there are 8,000 possible tripeptides (peptides built of three amino acids), 160,000 possible tetrapeptides (peptides built from four amino acids), and a staggering 3.2 million possible pentapeptides of the size that Merrifield was working on in his postdoctoral research. As the number of amino acids in the peptide grows by one, the number of possible natural peptides increases by a factor of 20.

Furka realized that there wouldn't be enough matter in the universe to make all possible proteins of even a modest size. Thinking about this problem stimulated him to think about how to make a large number of peptides. In 1980, when he was still pondering this issue, peptide chemists and biochemists were making peptides one at a time. Even with the accelerated pace provided by Merrifield's solid-phase synthesis methods, Furka calculated it would take 1,800 years to make all of the possible 160,000 tetrapeptides four amino acids in length, for example. A better method was needed.

By 1982, Furka developed a conceptual breakthrough for making large numbers of peptides at once. He wrote down a description of the method and had it notarized to legitimize it.[25] What Furka had conceptualized was the split-pool method of creating peptides (see Plate 4). It is an almost trivial idea, but one that is simultaneously powerful. The idea is to split and recombine plastic beads during the synthesis of peptides. To take a simple example, suppose you split a set of plastic beads into three flasks and that you attach a different amino acid to the beads in each flask. In other words, you add the first amino acid into the first flask so it will attach to all of the beads in that flask. You add the second amino acid into the second flask, and it will couple to all of the beads in that flask. Finally, you add the third amino acid into the third flask, where it will couple to all of the beads in that flask. At this point, you have three different flasks, each of which has a set of plastic beads attached to a different type of amino acid. All of the beads in a particular flask are exactly the same, it is important to realize, and no mixtures have been created. Thus, in this first part of the synthesis, you have carried out the "split" aspect of the split-pool method.

In the second part of the synthesis, you now combine all of the beads together into one flask. This is the "pool" aspect of the split-pool method. After pooling the beads, you have one large flask containing three types of beads (i.e., each bead has one of three possible amino acids on it, but no individual bead has more than one kind of amino acid on it).

Next, you again split the resin into three flasks. Then you couple a different amino acid to the beads in each flask (i.e., amino acid "one" into the first flask, amino acid "two" into the second flask, and amino acid "three" into the third flask).

At the end of this operation you have made nine different dipeptides. The three amino acids from the first round react with each of the three amino acids from the second round, yielding the full combinatorial assortment of possible dipeptides (i.e., 3×3=9). The crucial insight of Furka was that you have made *nine* different peptides, but you have only done *six* different reactions, three in the first round and three in the second round (i.e., 3+3=6). Thus you can make nine peptides by performing six reactions. That reveals the power of combinatorial peptide synthesis.

This ability to make nine peptides by performing just six reactions may not seem particularly powerful. But if you scale this approach to larger numbers, it becomes exceedingly effective. For example, by splitting the plastic beads into 20 flasks in each step, and by doing four rounds of amino acid couplings (including the first round, which is the attachment to the resin), it is possible to make the 160,000 tetrapeptides that Furka worried about originally. Moreover, instead of taking an estimated 1,800 years, this can be accomplished in only a few days and by performing just 80 reactions.

After Furka notarized his idea of split-pool synthesis, he began research on it, trying to demonstrate its feasibility.[26] His graduate student Mamo Asgedom was able to implement the methodology, and described it in his PhD thesis in 1987. The next year, in 1988, Furka presented the results at two international meetings, to little notice. Like Mendel, Furka was ignored, at least initially.

Then, in 1991, Kit Lam and colleagues published a paper in the widely read journal *Nature*, describing the same approach Furka had been pursuing for a decade in obscurity. Perhaps because of the wide readership of the journal, and Lam's position at a major medical center in the United States, the College of Medicine at the University of Arizona, the technique became widely known and associated with Lam. Only later did Furka's work come to light.

There have been bitter feelings between Lam and Furka. In a retrospective series of articles on the history of this field of combinatorial chemistry, published in 1999, Furka complained that he had given a presentation at the University of Arizona Cancer Center, where Kit Lam was located on April 1 1991, but no reference was made to Furka's prior work in Lam's subsequent 1991 paper in *Nature*. The implication of this is that Lam knew of Furka's prior work but suppressed mentioning it, perhaps fearing it would jeopardize acceptance of the *Nature* paper. Furka says he complained to *Nature*, and they promised a correction that never came.[27]

For his part, Lam has said that he had his eureka moment involving split-pool synthesis in 1989 as an Assistant Professor at the University of Arizona Cancer Center, and that he was unaware of Furka's 1988 abstract on the topic. By 1991, Lam says, he had implemented the method experimentally, and he presented it at an American Peptide Symposium in June 1991; it was published in *Nature* the same year. Lam claims that Furka did not clearly articulate the essential feature of split-pool synthesis—that each bead contains one unique compound—in his 1988 and 1989 abstracts or in his 1991 paper in *International Journal of Peptide and Protein Research*. Ultimately, it is fair to say that both Lam and Furka were instrumental in bringing this powerful new technology to the world of synthesis and drug discovery.

These methods enabled the chemical synthesis of millions of peptides using an efficient operational procedure. Other methods were developed at the same time to enable cells and small viruses to produce large collections of peptides, in an approach known as "phage display." The creation and testing of all of these peptide libraries began to take off as a new tool in science, and as a new industry.

In 1991 Stephen Fodor and colleagues at the Affymax Research Institute reported in the journal *Science* a new approach to making peptides and oligonucleotides, advancing the parallel synthesis method dramatically. They developed a way of building molecules in defined positions on a microchip using light-directed synthesis (see Figure 10.2). The strategy was to attach a chemical protecting group that would prevent each position on the chip from reacting until the protecting group was removed by exposure to light. They covered the chip with a mask that blocked some positions from light exposure, and allowed other positions to be exposed to light. In this way they were able to control which positions would be reactive at which times. By using a series of masks and coupling reactions, Fodor was able to program the synthesis of millions of oligonucleotides or peptides in a tiny space on a silicon chip. This became the basis of the successful company Affymetrix, which sells these chips for measuring the concentrations of thousands of messenger RNA molecules in cells, which in turn reveals which genes are on or off under a particular set of conditions.

Despite these profound advancements in synthesizing collections of molecules, a major stumbling block remained. Almost all of the molecules synthesized by these methods were peptides or oligonucleotides. Although these molecules are useful for understanding biological processes, they are not themselves likely to be drugs. The challenge was to apply these powerful new methods for constructing libraries to drug-like small molecules. Indeed, this turned out to

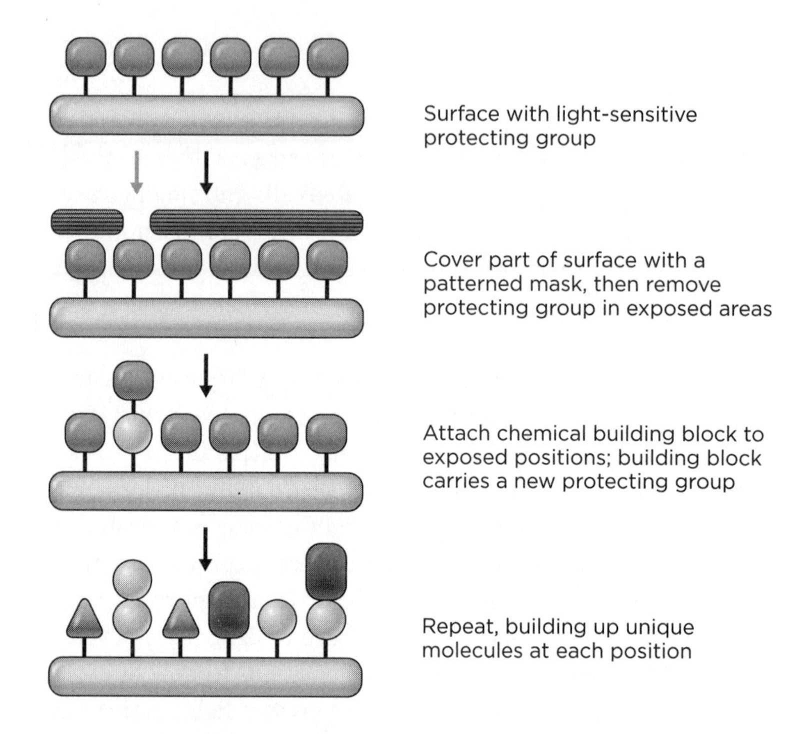

Figure 10.2 Light-directed synthesis. Molecules can be synthesized on a surface using light patterning, according to a method devised by Stephen Fodor and colleagues. This method is now used by Affymetrix to make gene chips for measuring the concentration of thousands of messenger RNA molecules in cells.

be a difficult challenge, because the methods that work well for peptides and oligonucleotides don't work for small molecules. A completely new approach would be needed to apply this technology to the problem of discovering new and better drugs—the kinds of molecules that might target undruggable proteins.

11

A VAST ARRAY OF DRUG CANDIDATES

One hope for tackling the undruggable proteins has been that by making and testing massive collections of small molecules, it would be possible to find drugs that can bind to these challenging proteins. However, prior to Merrifield's invention of solid-phase synthesis and the birth of combinatorial chemistry, this goal of massive screening of small molecules was impossible to realize. With these technologies in hand, it became possible to make large collections of molecules. The key question was: What kinds of molecules should be constructed?

The answer has its origins in Croatia in 1907. It was in that year that Leo Sternbach was born as the child of a pharmacist. Sternbach grew up to study chemistry and pharmaceutical science in Cracow, Poland at Jagellonian University and then worked at the elite ETH in Zurich, Switzerland (the Swiss Federal Institute of Technology). Before long, Sternbach moved to the private sector, joining the Swiss pharmaceutical company Hoffmann-La Roche, transplanting himself to Roche's facility in New Jersey.[1]

FINDING A DIAMOND IN THE ROUGH

Medicinal chemists like Sternbach focused on synthesizing a related set of small molecules with a particular chemical backbone, from among a virtually infinite number of possibilities. Sternbach was working with a group at Roche that was interested in developing new tranquilizers, or mood-altering drugs. The earlier generations of tranquilizing agents had serious defects. For example, the previously popular barbiturates such as phenobarbital (FEEN-oh-bar-bih-TAL) were addictive and potentially dangerous if taken in excess.[2] Sternbach and his Roche colleagues decided to try to discover a new class of tranquilizers that might have superior properties compared to the existing agents.

As a postdoctoral researcher in Cracow, Poland, Sternbach had attempted to synthesize compounds called benzheptoxdiazines (BENZ-hep-tox-DIE-ah-zeens). Chemical names are systematic, and they succinctly describe the chemical structure of a drug molecule. In this case the *benz-* prefix indicates the presence of a flat ring with six carbon atoms, the *hept-* root indicates the presence of an additional ring with seven atoms, the ox- root indicates that one of the seven atoms in this ring is an oxygen atom (instead of the usual carbon atom), and the *diazine* root indicates that two nitrogen atoms are present in this 7-atom ring (*di-* means two, and *az-* means nitrogen). Thus, despite the long name, these compounds were relatively compact, consisting of just two rings fused to each other, one having 6 carbons, and the other having seven atoms, composed of four carbons, two nitrogens, and an oxygen. To visualize what these molecules look like, imagine attaching a heptagon (7-sided regular polygon) to a circle.

These benzheptoxdiazines were originally explored in 1891 by Karl von Auwers and Friedrich von Meyenburg but had hardly been studied again until Sternbach's 1933 postdoctoral work.[3] Based on apparently little rationale, Sternbach and his colleagues began synthesizing a series of these compounds to study their biological activity.

Why would Sternbach choose to make these particular compounds, among the vast number of possible small molecules that could be made? There was no known biological activity for these compounds, and no compelling reason for choosing to make them. This raises an interesting general question: How do chemists choose what molecules to make when searching for new drugs? In one sense this is a trivial question, as some researchers argue that small molecules are largely interchangeable and what matters is not so much the particular structures you choose to make and test, but rather how *many* you choose to test.

Although this philosophy is not frequently articulated as such, it is the underlying premise in many modern screening approaches. In other words, to find a compound that inhibits a specific protein, a researcher will amass a large collection of somewhat arbitrarily chosen small molecules for testing. In almost all such modern research programs, the likelihood of success is justified based on the number of molecules to be tested, rather than on their unique chemical structures. In other words, researchers think it is more likely that a screening campaign will be successful if 100,000 compounds are tested than if 10,000 compounds are tested, and little thought is usually given to the structures of the compounds, assuming they are all relatively drug-like small molecules.

Sternbach's choice to make benzheptoxdiazines was based largely on the fact that he knew how to make these compounds (from his early postdoctoral work) and that they had not previously been explored as drugs.[4] At the time, synthetic methods were less developed than they are today, so synthetic accessibility was an important concern.

However, after trying to make a number of these compounds, Sternbach realized that the structures of the compounds he did make were not what he had thought they were. This highlights one of the challenges of synthetic chemistry. A chemist mixes two compounds together in a flask, adds additional reagents, and hopes to obtain a particular product out of the reaction. In order to determine if the reaction has made the desired product, the chemist needs to do a series of tests on the product to determine its chemical structure. Unfortunately, sometimes one can be misled by these tests, especially if the molecule produced is an isomer of the desired one, that is, if the product has the same atoms as the desired compound, but connected together in a different way.

Perhaps because of these complications, Sternbach abandoned the synthesis of these compounds (there was little rationale supporting their synthesis in any case). He bottled up the few compounds he had made, even though the correct structures were not known, and placed them on a shelf for storage. Two years later, in a laboratory cleanup, he was disposing of old materials in the lab, and came across one of these stored compounds, which he had labeled Ro-5-0690. Without much thought, since the compound seemed relatively pure, he sent it to be tested in the tranquilizer assays running at Roche. He expected to terminate his study of these compounds and switch to more promising compounds; Sternbach ordered this one last experiment to finish the project.

Sternbach was informed that the unknown compound tested had striking activity that was superior to the existing marketed tranquilizers. Based on this surprising and exciting result, Sternbach vigorously pursued the structural determination of this compound—what exactly had he made? A series of spectroscopic analyses and chemical degradation studies led him to conclude that this molecule was actually a 1,4-benzodiazepine (BEN-zoh-di-AYZ-ah-peen), a new class of compounds (see Figure 11.1).

This new compound proceeded rapidly into clinical testing for its anti-anxiety effects. Clinical investigators were enthusiastic based on the initial dramatic results obtained, and thousands of patients were soon being treated with the drug. In 1960 the drug was approved for sale under the name Librium, just three years after the discarded compound of unknown composition was discovered in the lab cleanup. Given that typical drug development timelines in the present era are 10 to 15 years, this was a breathtaking success.

However, there were suboptimal characteristics of Librium. It tasted bitter and was somewhat unstable. In an attempt to improve

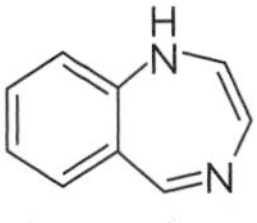

Core benzodiazepine structure

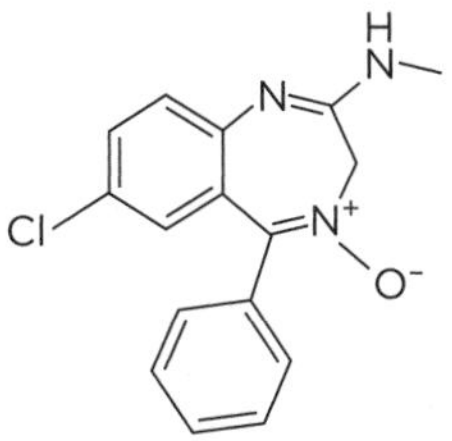

Structure of Librium, the first benzodiazepine drug

Figure 11.1 Benzodiazepine structure. The core structure of a benzodiazepine is shown, along with the structure of the first benzodiazepine drug, Librium.

upon Librium, Sternbach and his colleagues tried to determine the features that made it unstable, and experimented with new, related compounds that were more stable and less bitter. They succeeded in creating an even more potent, stable, and effective compound that was introduced in 1963 under the name Valium (with the generic name diazepam).

Valium grew extremely popular for a time as a treatment for anxiety, as well as other conditions, such as insomnia, seizures, and muscle spasms. Ironically, because of its benign safety profile compared to earlier drugs, Valium became widely abused and overprescribed. It soon was nicknamed "Executive Excedrin" because of its ability to relax executives coping with stress, and "Mother's Little Helper," based on a Rolling Stones song about anxious mothers using Valium. The potential for abuse eventually led to its decline in popularity.

In the interim other pharmaceutical companies realized the potential for this class of compounds, and additional benzodiazepines were introduced into the marketplace in 1965 (Serax), 1970 (Dalmane), 1972 (Tranxene), 1975 (Clonopin), 1977 (Ativan), and 1977 (Verstran).[5] For the period of time from 1969 to 1982, however, Valium was the single best-selling drug in the United States. Even today, it is a useful medicine to treat a variety of conditions, when used appropriately.

A PRIVILEGED SCAFFOLD

Sternbach had discovered a small molecule structure—the benzodiazepine scaffold—that had dramatic clinical utility in terms of having a tranquilizing effect. This alone was important and had a lasting impact on society and medicine. But, in time, the benzodiazepine scaffold would come to represent something even larger and more important than this.

In 1986 Ben Evans and his colleagues at Merck Sharp and Dohme Research Laboratories reported in the *Proceedings of the National Academy of Sciences* the discovery of a compound that could block a crucial protein-protein interaction involved in the formation of ulcers and gastrointestinal cancers.[6] This had never been done with a drug-like small molecule before. Surprisingly, this feat was accomplished with a benzodiazepine—the same type of molecule that reduced anxiety in patients.

Cholecystokinen (KOH-leh-sis-toh-KAI-nen), abbreviated as CCK, is a peptide hormone that controls secretion of material from the gallbladder and pancreas. Drugs that block the binding of CCK to its receptor protein on cells were thought to be promising candidates to treat ulcers, cancers, and possibly other conditions. However, development of CCK antagonists was thought to be difficult to impossible, because blocking the interaction between a peptide hormone and its receptor was a potentially insurmountable challenge, as most protein-protein interactions were assumed to be intractable. Although peptides could be found to block such interactions, these, as we have seen, are not suitable for use as drugs, and so cannot be developed into medicines. Small molecules, on the other hand, were thought to be too simple to block such a complex interaction as the association of a peptide with its receptor protein.

Evans and his colleagues approached this problem with a creative eye. Prior to their work, a complex natural product, asperlicin (as-per-LEY-sin), had been found to weakly block the interaction of CCK with its receptor. Although this was potentially a useful observation, asperlicin was a large, complex molecule and was unsuitable as a drug candidate because it couldn't be administered orally, had poor solubility in water, and was only a modestly effective blocker of CCK.

Evans examined the chemical structure of asperlicin in an attempt to come up with a smaller drug-like molecule that could block the binding of CCK to its receptor.[7] Quite surprisingly, he found that embedded within the large natural product structure of asperlicin was a benzodiazepine structure. He knew of Sternbach's work, and knew that benzodiazepines were the kinds of molecules that could become drugs. Therefore, Evans and his colleagues designed new molecules based on the benzodiazepine part of the asperlicin structure, and with suitable modification, created an effective small molecule benzodiazepine inhibitor of the CCK receptor. This was a significant breakthrough in finding drug-like molecules that could block the interaction of peptide hormones with their receptors.

In a subsequent paper in 1988, Evans and his colleagues further optimized their benzodiazepine CCK receptor antagonists.[8] These authors commented on the striking and unexpected observation that some types of benzodiazepines could relieve anxiety by binding to a protein in the central nervous system and other benzodiazepines could bind to the CCK receptor. Different benzodiazepines had these different activities, because they were decorated with different chemical appendages, but the core skeleton of the molecules was the same. Evans suspected there was something special about the particular arrangement of atoms that forms a benzodiazepine structure, allowing it to potentially interact with many different proteins.

Evans and his colleagues noted that another benzodiazepine had been found that binds to the opioid receptor, which is the same receptor that binds morphine (discussed in Chapter 2). In their discussion Evans and his colleagues wrote, "Thus, a single ring system . . . provides ligands for a surprisingly diverse collection of receptors, the natural ligands for which appear to bear little resemblance to one another or to the benzodiazepines in question."[9] They further speculated: "These structures appear to contain common features which facilitate binding to various proteinaceous receptor surfaces, perhaps through binding elements different from those employed for binding of the natural ligands."

Finally, in the concluding paragraph, they ended with this statement: "What is clear is that certain 'privileged structures' are capable of providing useful ligands for more than one receptor and that judicious modifications of such structures could be a viable alternative in the search for new receptor agonists and antagonists." Thus Evans and his colleagues introduced a key concept in the search for small molecule drugs—the concept of the *privileged structure*. These privileged structures, such as the benzodiazepine structure, are capable of binding productively to multiple different proteins in cells, and therefore represent an efficient way to find small molecules that bind to proteins. Depending on how the core benzodiazepine structure is decorated, it can lead to compounds with the ability to bind to a surprisingly diverse array of proteins.

A LIBRARY OF PRIVILEGED DRUG-LIKE STRUCTURES

Evans's work would have a profound influence on the research direction of a young chemistry student named Jonathan Ellman. Ellman did his graduate research in the laboratory of David Evans at Harvard University from 1984 to 1989, where he worked on making each of the two possible mirror images, called enantiomers (en-AN-tee-oh-murz), of amino acids and modified peptides. In the course of his doctoral work he became quite familiar with the solid-phase method of Merrifield. He read about the pioneering efforts of those who were trying to adapt solid-phase synthesis to the production of peptide libraries.

After being awarded his doctorate in chemistry, Ellman did postdoctoral research with Peter Schultz, one of the leaders in the research field at the interface of chemistry and biology. Schultz had recently started as a new faculty member in the chemistry department at the University of California at Berkeley. His lab and his professional stature grew rapidly, and he eventually moved to the Scripps

Research Institute in La Jolla, California. Along the way, he also became director of the more than 500 people working at the Genomics Institute of the Novartis Research Foundation.

In a short time Schultz accomplished impressive and diverse feats, including creating a combinatorial materials technology that could rapidly screen for materials with new properties, engineering the first bacteria with 21 amino acids instead of the usual 20, and converting antibodies, which normally do not have the ability to catalyze chemical reactions, into enzymes, which do catalyze chemical reactions. He also became a leader in constructing and testing collections of small molecules, both for drug leads and as tools to understand biological processes. He founded a large number of companies including Affymax, which was acquired by Glaxo for their combinatorial chemistry technology, and also led to the creation of Affymetrix, one of the key companies that make genechips for measuring the concentrations of messenger RNA molecules in cells and tissues.

Ellman came to the Schultz lab before many of these achievements had been realized, when there was great potential and a ferment of ideas. Ellman became fascinated by the power of selection methods. He learned about generating and testing huge collections of antibodies and peptides for new functions; for example, antibodies were selected that could catalyze a specific chemical reaction of interest, just as a natural enzyme does.

Ellman read the papers of Ben Evans. This catalyzed his thinking about what research direction to take as he set up his own research laboratory. He was intrigued by the suggestion that benzodiazepines represented a privileged structure that would be efficient for finding new compounds that bind proteins of interest. He furthermore thought that if he could apply the solid-phase synthesis methods that were used for making peptide libraries to the synthesis of small molecules, such as benzodiazepines, it would lead to useful compounds and potential drug leads.

Ellman initiated his new laboratory with this project. This was ambitious, as it was an attempt to reengineer solid-phase chemistry to make it compatible with the huge range of reaction conditions needed to make small molecules such as benzodiazepines, instead of the limited and well-worked-out conditions needed to make peptides. Another key difference between the solid-phase synthesis of peptides and benzodiazepines is that peptides are polymers. That is, they are repeating sequences of the same building blocks (amino acids). Benzodiazepines are not polymers, but are assembled from simpler materials, put together under the right reaction conditions.

Ellman, working with his first graduate student Barry Bunin, succeeded in adapting this different chemistry to the solid-phase environment, and created 10 different benzodiazepines on solid phase.[10] Although this was a small number of compounds, it represented a proof of concept for this approach. The Ellman lab proceeded to make larger collections of benzodiazepines and tested them for a variety of biological activities. They worked with Gary Glick to test some of these new molecules. Glick is at the Ann Arbor campus of the University of Michigan, where he is a member of the Chemistry Department and the Department of Biological Chemistry and is Director of the Chemical Biology program.

One of the benzodiazepine molecules made by Ellman's lab and tested by Gary Glick and his colleagues was found to prevent lupus in mice; lupus is an autoimmune disease caused by an overactive immune system, resulting in excessive inflammation in multiple organs and tissues. This benzodiazepine was found to bind to the mitochondrial adenosine triphosphate (ATP) synthase protein complex and cause production of reactive superoxide molecules. Superoxide is derived from oxygen found in the atmosphere (i.e., O_2) by addition of a single electron. This reactive molecule can cause damage to cell components and induce cell death. In this case the end result was the selective death of activated immune cells, leading to amelioration of the lupus disease process in mice.[11] Thus one of these benzodiazepines created by Ellman and his coworkers led to a compound that could bind to yet a different protein in cells—the mitochondrial ATP synthase. This was further evidence that the benzodiazepine scaffold was a privileged and productive one.

Later, Glick founded a company, Lycera, to develop this benzodiazepine as a therapeutic agent for lupus and possibly other autoimmune diseases. The company was founded in 2006 and raised nearly $50 million dollars from investors to develop their drug candidates into drugs. However, more important than the finding of new benzodiazepine drug candidates was the demonstration that small molecule libraries could indeed be created on solid-phase resins, using the efficiency and power of the solid-phase peptide synthesis method. This would open up the Merrifield approach to small molecule drugs.

THE PROBLEM OF TAGGING

Ellman's paper stimulated a new field of chemistry involving the synthesis of collections of small molecules. Before long, many chemists were creating solid-phase libraries of small molecules.[12] Although each chemical reaction needed to be adapted to solid-phase chemistry in a painstaking way, it was generally true that nearly any type of small molecule could be constructed on solid phase. However, the libraries constructed were small for a practical reason—it was not possible to keep track of where the compounds were. This became known as the problem of tagging.

When you synthesize a library of compounds using split-pool synthesis, a collection of plastic beads is split into different flasks in each step, and a different reaction is performed in each flask. Then, all of the beads are combined and the pool is split into a new series of flasks, where every type of bead is now present in every flask. This splitting and pooling is what allows the full combinatorial assortment of possible products to be made, where every compound made in the first step is reacted with every reagent in the second step.

However, if you think about carrying out this type of splitting and pooling operation on tiny plastic beads (typically less than half a millimeter in diameter), you realize that there is no way of keeping track of which compound is on which bead. Although it is true that each bead contains only one type of compound, in practice there is no way to know which bead has which compound. Although the same problem is manifest whether you are making peptides or small molecules, the problem can be circumvented in the case of peptides. This is because there are good methods for detecting tiny amounts of peptides to figure out after the fact which peptide is on a particular bead. In other words, you can make a million peptides via split-pool synthesis and then test the peptides for protein-binding activity while they are still attached to beads. Those beads that bind the protein of interest can be analyzed to determine the identity of the peptide on each bead.

Unfortunately, this can't be done for small molecules, because unlike peptides, there is no general way of figuring out the structure of small molecules. There is an amusing scene in the 1992 movie *Medicine Man*, with Sean Connery, in which Connery finds a rare natural product in the jungle and injects it into a sophisticated (but portable) chromatograph. Immediately, the chemical structure of the molecule pops up on the screen. Unfortunately, there is no machine that can do this. Instead, the chemical structure needs to be worked out using a variety of complicated tests and human reasoning, and using substantial amounts of material—more than can be obtained from one small bead.

Thus, in trying to construct massive small molecule collections by split-pool synthesis, a major stumbling block was figuring out how to know which compound is on each bead. This problem was solved by W. Clark Still.

Still did postdoctoral training at Columbia with the synthetic chemist Gilbert Stork and then took a job as an assistant professor at Vanderbilt University in Nashville. Within two years he was recruited back to Columbia, where he remained for 21 years before retiring. Still became interested in the tagging problem. The solution was to add a set of tag molecules onto each of the beads in each flask that would uniquely specify what reaction had occurred in that flask. Later, when the tags were detected, the identity of the tags would tell you which reaction had occurred on each bead, and it would be possible to infer what the structure of the compound on each bead is.

Although this concept was straightforward in principle, it was challenging to implement. The tags needed to be highly reactive, so as to be able to attach to the plastic beads reliably. But the tags would then have to be completely inert, so as not to be affected by any subsequent reactions performed on the beads. Finally, there would have to be a simple, reliable way to liberate the tags from the beads and decode their identity, thus allowing one to infer the structure of the small molecule present on each bead.

Still came up with an elegant solution to all of these problems in the form of carbene tags.[13] These tags have a chemical functionality—the carbene—that allows them to attach reliably into the plastic material of the bead itself. Still incorporated a functionality that allowed them to be reliably released from the plastic material at the time of reading the tag. The tags could then be easily detected with high sensitivity on a special type of instrument called an electron capture gas chromatograph. Thus Still was able to solve the longstanding tagging problem.

MAKING LIBRARIES OF SOPHISTICATED COMPOUNDS

Still patented his tagging invention and licensed it to a start-up company named Pharmacopeia that focused on creating large combinatorial libraries for drug discovery. This new company was successful and arranged a number of collaborations with large pharmaceutical companies. Before long, Pharmacopeia had a successful initial public offering, in which they were able to raise $43.8 million in funds from public investors. At one point, the market capitalization of the company was in excess of $1 billion, as the promise of more efficient drug discovery through combinatorial libraries loomed over the industry. Pharmacopeia, through the Clark Still patents, had a lock on the technology needed to make huge small molecule libraries using split-pool synthesis, and this was anticipated to give them a huge competitive advantage in discovering new drugs.

However, there was a sense in time that the promise was not fulfilled, and excitement over Pharmacopeia faded by the turn of the millennium. The company survived for some time, but in 2008 they were acquired by another company, Ligand Pharmaceuticals, for just $70 million.

What happened to Pharmacopeia and to combinatorial chemistry in general? As with Pharmacopeia, the field of combinatorial chemistry seemed to fade in promise, as the initial hope for a dramatic improvement in drug discovery efficiency was not realized. There is still debate about this point, as many active researchers in the area of combinatorial chemistry bristle at the notion that the field has failed to deliver on its initial promise. I recall attending a combinatorial chemistry meeting in 2005 where one of the speakers commented on this issue. He noted that every major pharmaceutical company had incorporated combinatorial methods into the drug discovery process, and it was now a routine aspect of the business. In what sense could the field be perceived as a failure?

While it is certainly true that combinatorial chemistry is now widely used in the pharmaceutical industry, it is largely for a different purpose than originally intended. The original goal was to make huge libraries for discovering small molecules that could bind to new proteins. That application mostly has not been realized. However, combinatorial methods did turn out to be efficient for optimizing compounds that have already been discovered. In this case the goal is to make a small number (a few hundred at most) of related structures and test their biological activity. This is routinely done in academia and in industry. In addition, these small libraries are added to large screening collections, leading to the possibility that compounds synthesized for one project will turn out to be good starting points in a future project.

What was the problem with making large libraries for screening? One issue that arose was that the compounds that were synthesized were usually simple and readily accessible using basic chemistry.

These simplistic libraries were created because such reactions were the easiest to adapt onto solid phase, and because they were the most tolerant of a wide range of substrates, ensuring that the same reaction would work in every flask, irrespective of the specific building blocks present in each flask.

However, not a great deal of consideration was given to which compounds should be made from an ideal perspective, and as a result, many of these compounds produced did not have properties compatible with advancement as drug candidates. For example, they were often too large, too insoluble, or too unstable to be developed as drugs.

The largely unmet challenge was to take more complicated chemistries that could be used to synthesize molecules with the complexities of natural products and adapt these methods to solid phase. However, such a goal would require the long-term commitment of highly skilled synthetic chemists. Traditional synthetic chemists, since the time of Merrifield, had viewed solid-phase and combinatorial methods with skepticism, if not disdain. Thus a cultural gap prevented the union of state-of-the-art synthetic methods with combinatorial technologies.

Into this gap walked my doctoral thesis advisor, Stuart Schreiber. Stuart grew up in Virginia, and did his doctoral work at Harvard University in synthetic chemistry with the great chemist R. B. Woodward. Sadly, Woodward passed away during Stuart's graduate work, but he completed his thesis under the supervision of Yoshi Kishi in the same department. Instead of taking the normal route of doing postdoctoral training, Stuart began directly as a professor at Yale University in 1981. Within three years he received tenure at Yale, and in 1988 he was successfully recruited to the Chemistry Department at Harvard.

Stuart's initial independent research program involved the total synthesis of complex natural products, which is the type of cutting-edge synthetic chemistry that garners the respect of the traditional chemistry community. He quickly developed into a leading figure in this field. However, instead of continuing to synthesize individual natural products, he became fascinated with the challenge of efficiently discovering small molecule modulators of protein function. Just as the field of combinatorial chemistry was getting under way, he saw the potential value in combining state-of-the-art synthetic chemistry with combinatorial methods.

Stuart's first major foray into this arena occurred while my own graduate studies were under way on a different topic in his lab. Sitting next to me in the laboratory was a fellow graduate student, Derek Tan, as well as my future CombinatoRx cofounder Mike Foley. Derek and Mike undertook with Stuart and a postdoctoral scientist named Matthew Shair to build a large library of small molecules, but with the complexity of natural products. The hope was that this would represent a quantum leap in the usefulness of these large libraries, and would solve the issues that plagued the ubiquitous simpler libraries.

They succeeded in making a collection of over 2 million compounds with structural features similar to complex natural products. The rationale was that natural products often have powerful biological activities—but they are also difficult to synthesize because they often have complicated architectures. By making many compounds with structural features similar to natural products, Stuart and his colleagues believed they could find powerfully active molecules that had never been made by nature. They published the first collection in 1998 in the *Journal of the American Chemical Society*, and it was cited by the journal *Science* as one of the top ten scientific breakthroughs of the year.[14] I remember looking at the collection of compounds in a plastic tube that could fit comfortably in the palm of my hand, and being awed by the thought that two million different complex compounds were present in that tube.

Despite the impressive size and nature of that library, we had trouble making good use of it. Once the library was available, I worked with Derek and Mike to try screening the compounds in different assays that I was interested in. Our initial pilot screen worked well. In this first experiment, we tested 500 of these new compounds for their ability to turn on a reporter gene that I had been using. This gene was known to be turned on by the transforming growth factor beta signaling pathway in mammalian cells, which is a pathway that becomes defective in many tumors. Therefore, finding compounds that could turn on this pathway was of interest because they might become therapeutic agents, and they might illuminate new aspects of this signaling pathway.

Out of the 500 initial compounds tested, we found six structurally related compounds that were capable of turning on this reporter gene. Although the extent of activation was modest, these results demonstrated that at least some of the two million new compounds possessed biological activity, which was a finding that greatly excited all of us. At this point, we tried to find a way to test a larger number of these compounds to find one with more striking activity.

This turned out to be extremely tricky. The compounds had been synthesized on small Tentagel beads that were about one-tenth of a millimeter, or the width of a human hair, in diameter. Each tiny bead did not have a lot of material on it. There was only enough material on each bead to run one small experiment. To test the compounds, we needed to find a way to spread them out into individual test tubes or wells, release the compounds from the beads, transfer the compound solution into a new test tube or well, and then add cells and perform our assay.

We eventually worked out a way of testing tens of thousands of the compounds for their effects in cells, but the method was not reliable. When we identified a bead that supposedly had an active compound, we had to go back and cleave and identify the tags from the bead, resynthesize the compound (which would take several weeks), and retest it. Although we found some modest activities, many putatively active compounds could not be confirmed.

Thus, although this was potentially a transforming technology, there were technical hurdles that had to be overcome in order to make it practical. Stuart and his group members set out to solve these issues over the next few years. I completed my PhD on other topics around that time, and moved to the Whitehead Institute as a Fellow, to begin directing my own laboratory. I pursued other interests, and watched from afar as Derek, Mike, Stuart, and others in the laboratory tried to solve these technical challenges.

In time Stuart and his laboratory members were able to create a superior platform for making and testing complex, natural-product-like compounds using combinatorial methods. Much of this charge was led by Mike Foley. A few years later, Mike, Stuart, and others founded a new company, called Infinity Pharmaceuticals, to take advantage of their new platform. They created a series of sophisticated chemical libraries using this new platform, providing them with large quantities of each compound.

The new platform solved many of the issues we had faced in those pioneering experiments. Infinity was successful at industrializing this platform and signed a number of collaborative agreements to test these molecules against proteins that pharmaceutical companies were interested in making drugs against. Several of these programs have yielded drug candidates, and these are being developed by Infinity and their collaborators. Meanwhile, Stuart and his laboratory members spent the next decade pioneering this new field at the interface of synthetic chemistry and combinatorial technology. Stuart coined the name diversity-oriented synthesis (DOS) for this field, and developed a series of methods and approaches that allowed for the systematic construction of complex, diverse, small organic molecules using combinatorial technology.

One of the most exciting breakthroughs was developed by Stuart and his graduate student named Martin Burke, now a professor at the University of Illinois. They developed a way of synthesizing collections of small molecules with a new type of diversity—skeletal diversity. A typical library of peptides contains a series of molecules that all have the same skeleton, or molecular framework, connecting different building blocks. In the simplest case of a dipeptide, there are two positions for the building blocks, and one bond connecting them. In a library of all 400 possible dipeptides, every peptide has the same type of connection between the two amino acids. The only difference between the peptides is in the specific building blocks—the amino acids—that are found at each position.

What Stuart and Marty developed was a way to alter the connectivity of the building blocks in a systematic way during the construction of a library of compounds. I remember being enthralled when I read their 2003 paper in *Science*.[15] The compounds produced would no longer be required to have the same backbone connectivity, and would be more diverse in terms of their three-dimensional shapes. This concept, as well as other concepts pioneered by Stuart and his group, built this new field, DOS, into an exciting forefront in chemistry. There is now the possibility of realizing the original promise of the combinatorial approach for making large libraries of complex small molecules.

However, the field still faces that trivial, yet profound, question of what kinds of molecules should be made and tested, out of the virtually infinite number of possibilities. Some researchers are focusing on natural-product-like compounds, while others have focused on synthetic accessibility. It may be fruitful to return to the work of Ben Evans and the concept of privileged scaffolds. In 2010 I wrote a review article on this topic with Scott Snyder, a professor in the chemistry department at Columbia, and our joint graduate student Mathew Welsch.[16] In this paper we articulated the notion that privileged scaffolds may yet be the optimal approach for designing small molecules, since they have an increased probability of yielding active compounds in diverse biological tests.

Since the work of Evans and Ellman on benzodiazepines, other scaffolds have been shown to have a privileged nature. Incorporating these privileged scaffolds into chemical libraries could be productive for generating compounds that bind to undruggable proteins, if the privileged nature of these compounds extends to these intractable proteins. On the other hand, making and screening large and diverse libraries of complex molecules with novel architectures may be an effective means to discover new privileged structures.

It is too early to tell whether these new libraries will solve the problem of the undruggable proteins, or at least some subset of these proteins. In time, these new collections of compounds may address some undruggable targets. It may also become clear that additional tools are needed. There is every reason to be optimistic, but there is also a possibility that some targets will be completely resistant to small molecules of any kind.

What will researchers do in these cases? Will we turn away from these undruggable proteins, or will we find a way to target them without using small molecules? This would be a challenge indeed, given that small molecules represent the workhorse of the pharmaceutical industry for the last century. However, some recent work suggests that an entirely new class of drugs could be around the corner. To explore the potential of these molecules, we will need to step outside of the small molecule box.

12

MOVING OUTSIDE THE SMALL MOLECULE BOX

Every once in a while, an event occurs in science that shatters the existing paradigm for what is possible, and huge new vistas are suddenly opened. These changes can come remarkably suddenly, within the space of a few years. What was once unthinkable can become suddenly possible, or even routine. These are exciting moments in science and exhilarating to observe.

Yet, despite their rapidity, these transformative technologies are evolutionary in the sense that they steadily build one discovery upon another. For example, before Merrifield, peptide synthesis was arduous, and it was ludicrous to contemplate synthesizing millions of peptides in any time frame, let alone in a few weeks. Merrifield's work was revolutionary, yet it built upon previous efforts to create and use solid-phase resins. After Merrifield, other researchers expanded the utility of solid-phase synthesis from single peptides to libraries of peptides. In time the technique was broadened to small molecules, and was no longer limited to peptides. This was a gradual, evolutionary, but still transformative, technology that changed what was possible.

The problem with recognizing these transformative events as they happen, is that one can never tell if a potentially exciting breakthrough will stand the test of time, be built upon and change a scientific paradigm, or more commonly, will be less robust than thought at first and recede into the backwaters of scientific history. We are currently at a crossroads in the search for entirely new classes of drug molecules. These new types of molecules might allow drugs to be created against proteins that are completely resistant to small molecules; thus, it might become possible with this new approach to make drugs against currently undruggable proteins.

THE CHALLENGE OF BIOLOGIC DRUGS

Small molecules have been the foundation of the pharmaceutical industry for the last century for a simple reason. Many small molecules, unlike almost every other kind of molecule, can be administered in the form of a pill, and can be distributed throughout an organism without being blocked by cellular membranes or destroyed by conditions found in the gastrointestinal tract or in the bloodstream. Thus the long-standing wisdom has been that larger molecules, such as proteins, DNA, or RNA do not make good drugs.

This view changed beginning in the 1980s with the development of efficient methods to manufacture proteins and antibodies. These large molecules, referred to in the industry as *biologics*, have turned out to be effective drugs for specific indications; in particular, proteins can be used to treat diseases in which the drug needs to attach to a protein on the surface of cells, or to replace a missing protein that normally would circulate in the bloodstream.

For example, Amgen's erythropoietin (eh-REETH-roh-POIY-oh-tin) is a fabulously successful drug that replaces a natural protein normally found in human blood. This drug, abbreviated as EPO, stimulates the production of red blood cells. Patients with certain types of anemia can be productively treated with EPO. The drug has sales of several billion dollars each year, indicating it fills an important medical need.

Biologics, such as EPO, can be effective drugs by acting on proteins outside of cells. However, many diseases require that a drug be able to access the interior of cells, sometimes in difficult-to-reach tissues. Despite the successes with biologics, it has been clear that these large drugs are not effective at intervening inside of cells, such as against the mutant RAS proteins in cancer cells. There are two reasons for this pessimistic view. First, large molecule drugs cannot penetrate across the membrane that surrounds all cells, and second, large molecules such as proteins are often unstable, and decompose in the presence of protein-destroying enzymes known as proteases. Thus there has been a consensus that proteins, peptides, and antibodies will not be effective for targeting intracellular proteins, which include most cancer-causing proteins, as well as proteins that control many other diseases.

A crack in this dogma emerged through the study of two different biological systems—how fly embryos develop, and the properties of the human immunodeficiency virus (HIV). An unexpected breakthrough in one field coming from an unrelated field is a common occurrence in science and is well known to practicing scientists. This is one of the reasons that many scientists are passionate about the need to fund basic, curiosity-driven research. You literally never know where the most transformative technologies will come from. Moreover, this story illustrates the power of interdisciplinary projects and multidisciplinary teams—by exposing yourself to ideas from other fields, you might gain a critical new insight or technology that can solve a difficult problem that lies between traditional disciplines.

THE CURIOUS PROPERTY OF THE ANTENNAPEDIA PROTEIN

One strand of this revolution started in the field of developmental biology. Since the time of Morgan, fly geneticists have been interested in identifying the genes that direct the development of an adult fly from an embryo. This is one of the most profound questions in all of biology: How do you take a single uniform cell (the fertilized egg) and turn it into a complex, three-dimensional multicellular organism with all of the associated tissues and organs? It seems miraculous, and yet we know that the entire process is controlled by nothing more than molecules interacting with each other. Nonetheless, development is truly a wondrous process, as any parent can attest.

Fly geneticists try to understand the developmental process by looking for mutants that perturb it.[1] These mutant genes are the ones that control the process of creating an adult organism from a single cell embryo. By studying these controlling genes, it is hoped that researchers can understand the principles and molecular details of the entire development process.

Morgan himself had observed mutant flies with striking developmental changes. One of the most dramatic developmental mutants is a homeotic mutation, which is a mutation in a gene controlling how the basic body plan of an organism is established during development. Flies with these bizarre mutations have one body part that is transformed into a different body part. This alters the architecture of the fly. For example, *Drosophila* fruit flies normally have one pair of wings. In some homeotic mutants a second pair of wings appears where ordinary flies have no wings.[2] Another homeotic mutation converts the flies' antennae into an extra pair of legs.[3] The term homeotic mutant was coined by the British geneticist William Bateson in 1894, based on the Greek word *homeo*, which means "similar"; presumably, Bateson used the term because he was struck by the ability of these mutants to convert one segment of the fly into a similar structure from another segment. These homeotic genes are responsible for specifying the identity of each segment along the body; that is, whether it should form a head, wings, antennae, or another structure.

The advent of DNA technology allowed a number of groups to identify the genes responsible for these intriguing homeotic mutations. One such gene was named antennapedia (an-ten-ah-PEED-ee-ah), because it was responsible for converting antennae into legs (pedia).[4] The antennapedia protein was found to be a sequence-specific transcription factor—it functions by binding to a specific DNA sequence in the genome of the cell, and turning on expression of specific genes. Transcription factors are the "on" switches for genes, allowing these genes to be activated at specific times and places in an organism.

Antennapedia, like many other proteins, is divided into multiple domains, or functional modules. One crucial domain in the antennapedia protein is a 60-amino-acid section known as the homeodomain, which is responsible for binding to DNA. In a paper in *Proceedings of the National Academy of Sciences of the USA* (*PNAS*) in 1991, Alain Joliot and colleagues reported an unexpected property of the homeodomain.[5]

These researchers reported that they had synthesized the 60-amino-acid homeodomain in the laboratory, with the aim of introducing it into cells. This small domain does not represent the complete antennapedia protein, but has the capability of binding to DNA in the way that the full-length antennapedia protein binds to DNA. Joliot and colleagues expected that the homeodomain protein would block the normal antennapedia protein in cells from binding to DNA, and thus reveal the consequence of shutting off the real antennapedia protein. In other words, they were trying to test the effect of inactivating the antennapedia protein in cells.

This is just the sort of occasion where a small molecule inhibitor of the antennapedia protein would be valuable, since it could be added to the medium outside cells in culture, or even to intact flies, to determine the consequence of inhibiting antennapedia. Unfortunately, transcription factors such as antennapedia are considered undruggable, and no small molecule inhibitors are available for this sort of experiment. Thus Joliot and colleagues decided to chemically synthesize the homeodomain of antennapedia and introduce it into cells, thereby blocking the normal antennapedia present in these cells from functioning.

The homeodomain protein fragment was not expected to penetrate into cells, since it was known that proteins cannot penetrate across cell membranes. An attempt was made to introduce the protein into cells by mechanically agitating the cells, with the hope that this agitation would transiently disrupt the cell membrane and allow the homeodomain protein to enter the cells. This procedure was apparently effective, as the homeodomain protein fragment could be detected inside of cells after the agitation process.

In fact, the entry of the homeodomain protein into cells seemed almost too efficient. The researchers became suspicious that perhaps something unanticipated was happening. Thus they decided to do a control experiment. This is so crucial to the mind-set of scientists; it cannot be emphasized enough. It is not enough to do an experiment and get the desired result. You must think of all possible control experiments to rule out other explanations. Scientists recognize that no theory can ever be "proven." All we can do is disprove all of the possible alternative explanations we can think of until the only reasonable explanation remaining becomes the accepted theory.

I learned firsthand how crucial control experiments are. When I started my graduate work, I was trying to create a method for activating the transforming growth factor beta (TGF-beta) signaling pathway in cells. I spent two years working on this without success. I was excited when I finally found a way to do it. I did one last control experiment, which invalidated all of the conclusions I had previously drawn. I presented this result to Stuart Schreiber, my advisor. I half-jokingly said that I supposed I had done one too many control experiments. However, Stuart responded by sagely reminding me about a fundamental tenet of scientific research. He said, in effect, don't joke about that—our goal is not to publish papers, but to seek truth. It is far better to do the right controls and learn what is really going on than to rush to publish something that later has to be corrected. I have always valued this approach to science that he instilled in me—being right is the only thing that matters at the end of the day. In fact, once we figured out what was happening with this control experiment, it led to two papers instead of one.

In the case of the antennapedia protein, the control experiment would lead to a profound discovery, with significance far beyond the study of fly development. Joliot and colleagues did the control experiment of testing the effect of adding the homeodomain protein to cells growing in a dish without agitation, expecting nothing to happen. Unexpectedly, the homeodomain penetrated effortlessly across the cell membrane, becoming concentrated inside cells. They did an additional control experiment (to control for this surprising effect), testing whether a standard protein, ovalbumin, could penetrate into these same cells when tested alongside the homeodomain protein. As expected, ovalbumin could not penetrate into the cells. Thus there was a truly special property of the homeodomain protein that allowed it to slide right through the impenetrable membrane barrier of cells, entering into cells.

THE PHENOMENON OF PROTEIN TRANSDUCTION

A number of researchers began exploring this surprising new property of the antennapedia homeodomain protein. One of the first questions was, is the entire protein needed for this effect, or will a smaller piece of the protein suffice for penetrating across the cell membrane? Systematic analysis of mutants and fragments of the antennapedia homeodomain revealed that one part of the protein was sufficient for cell penetration. In fact, a peptide made up of just 16 amino acids was capable of penetrating into cells.[6]

What was the mechanism by which this small peptide could traverse the normally impenetrable cell membrane? Again, a systematic series of experiments illuminated aspects of the mechanism. Changing the sequence or three-dimensional shape of the peptide to disrupt its helical nature did not prevent cell entry. Thus the specific structure of the peptide was not being recognized by a protein on the surface of the target cells. This conclusion was consistent with the observation that there did not seem to be any cell specificity in the penetration effect—if a protein receptor for the peptide were involved, it would likely be present on some cells, but not others. On the other hand, changes to the amino acid sequence of the peptide could abolish the internalization, suggesting a specific mechanism of some sort was involved.

An unexpected observation, such as the ability of this peptide to cross the normally impenetrable cell membrane, can be a powerful route to new discoveries in science. Most of science moves forward using a hypothesis-based approach. In other words, researchers analyze the accumulated data in a field and formulate a hypothesis. Although this can be a powerful route to understanding existing phenomena, it is less effective at tapping into completely novel phenomena, which often come to light only through serendipitous observations. Crucially important is that the person observing the unusual phenomenon is able to recognize the importance and capitalize on it.

Once a serendipitous observation is made and confirmed, a hypothesis is needed to explain the mechanism of the observed phenomenon. In the case of the cell-penetrating peptides, the hypothesis that emerged was that these peptides were able to disrupt the structure of the cell membrane in a way that leads to internalization of the peptide. Specifically, it was suggested that the peptide causes a small ball-like structure (called a micelle) to form in the cell membrane, and this ball was the entry vehicle for the peptide. A number of experiments were consistent with this hypothesis. However, other explanations have been offered involving other cell-uptake mechanisms. The specific mechanism of cell penetration appears to vary depending on the peptide and what cargo is attached to it.

In addition to the antennapedia homeodomain peptide, a number of other naturally occurring peptides and proteins were shown to possess this unusual cell-penetrating property.[7] One of the other major proteins shown to have this cell-penetrating effect was the transactivating protein (called Tat) of the human immunodeficiency virus (HIV), which causes acquired immunodeficiency syndrome (AIDS). This phenomenon was observed by Frankel and Pabo as early as 1988, which predated the antennapedia homeodomain cell-penetrating discovery.[8]

Since the cell-penetrating effect was not limited to one peptide, it was of even greater importance, reflecting a more general property of some peptides. The exciting use of a cell-penetrating peptide would be to deliver molecules into cells that could not penetrate on their own. A number of researchers demonstrated that these cell-penetrating peptides indeed were capable of delivering pieces of DNA, peptides, or small proteins into cells. Thus a great number of possibilities became available. Using cell-penetrating peptides, it might be possible to deliver into cells large molecules, especially proteins and peptides, which were previously discounted as drug candidates because of their inability to access the interior of cells. Such delivery of proteins into cells became known as *protein transduction* and was soon a forefront area of research.

FROM CELLS TO ANIMALS

Although the use of cell-penetrating peptides in cell culture studies showed promise and generated excitement, the next several years were challenging, as researchers tried to test these new reagents in animal studies.[9] The first proof of concept of delivering a modified peptide to a mouse was shown by Ülo Langel's group in 1998, and delivery of a protein to a mouse was reported by Steven Dowdy's group in 1999.[10]

A number of other research groups continued to explore the potential therapeutic efficacy of cell-penetrating peptides in mice. In 2005 the research group led by Carlos Arteaga at the Vanderbilt University School of Medicine tested the utility of these peptides for delivering an antibody that could inhibit a key protein in tumor cells.

Antibodies were already known to be powerful reagents for targeting specific proteins on the surface of tumor cells. Antibodies are large proteins produced by the immune system for binding to and eliminating foreign proteins from the body. Herceptin, one of the most widely used antibodies for the treatment of breast cancers, is an example of a therapeutically effective antibody. Herceptin binds to the HER2/neu cancer-causing receptor on the surface of some breast cancer cells, blocking the cancer-causing activity of this protein and causing destruction of tumor cells. This leads to increased survival in a subset of breast cancer patients treated with Herceptin.

Antibodies can be effective because they have the capability of engaging in highly specific recognition of a single protein, and distinguishing one protein from all of the other proteins present in a cell or organism. However, there was not previously a viable strategy for using antibodies to target intracellular proteins.

Arteaga and his colleagues tried to address this deficiency using a cell-penetrating peptide.[11] They generated an antibody that could inhibit the three human AKT kinases, known as AKT1, AKT2, and AKT3. These proteins were known to be important in stimulating the survival of tumor cells in a variety of conditions, and contributing to tumor formation. Thus, if it were possible to inhibit these kinases in tumor cells, it might be possible to kill these tumor cells, especially if they had become addicted to the AKT kinases. A number of groups had been attempting to create small molecule inhibitors of the AKT kinases. This effort, while still promising, was difficult because of the challenge of creating specific kinase inhibitors, given the close similarity in shape of many human kinases, the success of imatinib notwithstanding.

Arteaga and colleagues were able to attach a cell-penetrating peptide to an antibody that inhibited the AKT kinases. They delivered this protein into tumor cells and blocked the activity of AKT kinases in these cells. Finally, injection of this antibody-peptide fusion into mice harboring tumors caused a reduction in tumor size. While this specific reagent still needs to be optimized for a number of properties, this represented a demonstration that antibodies might be effective drugs for targeting intracellular proteins; all that was needed was the attachment of a cell-penetrating peptide onto the antibody.

ARTIFICIAL PROTEINS

Naturally occurring cell-penetrating peptides are reasonably effective at penetrating into cells and delivering attached therapeutic molecules into the interior of the cell. However, researchers soon realized that there was no reason to limit themselves to the peptides made by nature. A number of laboratories, including those of Paul Wender at Stanford University and Shiroh Futaki at Kyoto University in Japan, sought to improve upon these naturally occurring peptides by making more effective cell-penetrating molecules.

Wender's most cited paper is his 2000 publication in *PNAS* on designing more effective cell-penetrating molecules—it has been cited more than 500 times in the last decade. In this paper he defined the chemical properties of the cell-penetrating peptides that allow them to traverse biological membranes.[12] Moreover, in the process of defining these properties he created a peptide that was more than 100 times better at penetrating into cells. This is a classic approach of chemists—they first synthesize a series of molecules that look like a molecule of interest, but vary slightly in their chemical structures. In doing so the chemists quickly learn which parts or properties of the molecule are important for activity, and which are dispensable. This leads to an understanding of the properties of the molecule that endow it with its biological activity. Along the way, this information can often be used to make more effective versions of the molecule. This is the process used to optimize drug candidates as they progress down the path to becoming actual drugs.

Wender discovered a key property of cell-penetrating peptides. He found that they should have suitably spaced positive charges. These positive charges needed to be in the proper context, specifically in the form of the chemical group found at the end of an arginine amino acid, called a guanidinium (gwan-ih-DIN-ee-um) group. The most effective peptides Wender created were repeating units of arginine.

Using this discovery about the requirements for cell penetration, Wender was able to design non-peptidic molecules with different backbones that also penetrated into cells exceptionally well. These designed molecules, and the polyarginine peptides Wender created, were all highly effective at penetrating into cells. Wender was able to attach these cell-penetrating molecules onto other peptides and transform them into potential drugs that could access the interior of cells.[13]

Numerous researchers then became energized about the possibility of improving upon cell-penetrating molecules. Why limit yourself to peptides, when chemists are capable of making almost any kind of molecule? A series of research groups proceeded to make non-peptide-based, guanidinium-containing molecules that could penetrate exceptionally well into cells.[14] Initially, the focus was on making molecular appendages that could be attached to proteins, facilitating entry into cells. Before long, a number of groups began exploring the possibility of making synthetic proteins.

Proteins are built from repeating units of amino acids. The structure of the connection from one unit to the next is invariant. The only thing that changes from one position to the next is the nature of the specific amino acid. To imagine what this looks like, picture a line of 100 people standing shoulder to shoulder. Imagine that each person in this long line has a hand extended out holding a different tool. One person is holding a hammer, while another holds a rake or a screwdriver. This is a repeating structure that varies only slightly in terms of what functionality is presented at each position.

A number of creative chemists began to ask: Are we limited to making proteins from these repeating amino acid units? What if we tried different kinds of molecules? Would repeating units of these other kinds of molecules form folded structures? These folded structures would no longer be considered proteins, because proteins are defined to be repeating units of alpha-amino acids, a specific type of amino acid. The more general term *foldamer* is used to refer to repeating units of other kinds of building blocks that fold up to form stably structured molecules.

To return to the analogy of the line of 100 people, this would be akin to making a line out of 100 chimpanzees, or even lions, zebras, or octopuses. What functionality could they project, and how would they behave when they were assembled in this way? This is a more general version of the protein-folding question.

The big question in protein folding has been this: How does a linear sequence of amino acids fold up into a unique three-dimensional structure? The more general question from a chemistry perspective is: How does a linear structure of any chemical type fold up into a unique three-dimensional structure?[15] Moreover, why did life choose amino acid polymers for proteins instead of some other kind of repeating unit? Is there something special, from a chemistry and biochemistry perspective, about repeating units of amino acids that is essential for life? Or could you design many other biochemistries of life? These are fundamental questions that need to be answered in order to be able to design artificial foldamers, and to understand why nature chose amino acids.

The foldamer name was coined in 1996 by Samuel H. Gellman. Gellman did his PhD work under the supervision of Ronald Breslow at Columbia University. Gellman then joined the faculty at the University of Wisconsin in Madison in 1987, where he has remained for his career.

In his 1996 paper in the *Journal of the American Chemical Society*, Gellman showed that he could create a folded polymeric structure that was not based on alpha amino acids, but rather was based on a type of beta-amino acid.[16] Beta amino acids have an extra carbon inserted into them, compared to alpha amino acids. In this study, Gellman and his colleagues showed that artificial polymers could be designed to fold into stable shapes. This was a first step towards creating artificial proteins with designed properties. Four years later, Gellman's group used this approach to create a foldamer that had striking anti-bacterial activity, presumably by forming a stable structure that could interact with and disrupt the bacterial cell membrane structure.[17] A number of other researchers were exploring similar strategies at the same time, including Dieter Seebach at the Federal Institute of Technology (ETH) in Switzerland and William DeGrado at the University of Pennsylvania.

Such researchers showed that it was possible to create artificial protein-like molecules with biological activity. This suggested it might become possible to create a new class of drugs, consisting of foldamers that could penetrate into cells and perform functions inside cells, such as replacing missing proteins, or inhibiting defective proteins. Indeed, a growing number of researchers have recognized the potential in this approach and have explored foldamer structures.[18]

In this vein, some groups have explored the idea of creating synthetic miniproteins. Alanna Schepartz, a professor at Yale University, has developed intriguing examples of miniproteins and foldamers. Schepartz grew up trying to be a chemist. She relates that as a child she used empty liquor bottles as reaction vessels to mix food substances. She went into science as a profession, earning her PhD from Columbia University, also working with Ronald Breslow. She did postdoctoral work with Peter Dervan at Caltech and started her own independent lab as an assistant professor at Yale, where she has remained for her career. She was the first female full professor in the physical sciences at Yale, paving a path for future woman scientists at Yale.

Schepartz's first miniprotein was reported in 2001, and numerous additional miniproteins with biological activities have been described since then. In 2007 Schepartz and her colleagues reported making a foldamer that could assemble in the way that proteins do, by having multiple foldamer molecules come together to form a complex of molecules.[19] Such productive interactions between molecules constitute one of the special features of proteins—Schepartz's lab showed that nonprotein, artificial polymeric molecules could engage in this kind of sophisticated intermolecular choreography. This suggested that artificial foldamers and miniproteins might ultimately accomplish sophisticated functions and be used to create drugs targeting otherwise undruggable proteins.

STAPLED PEPTIDES

Designed synthetic proteins will likely take some time to become approved drugs. However, a related, but simpler, approach is moving quickly towards clinical testing. This approach has been most notably developed by Gregory L. Verdine, a professor at Harvard University.

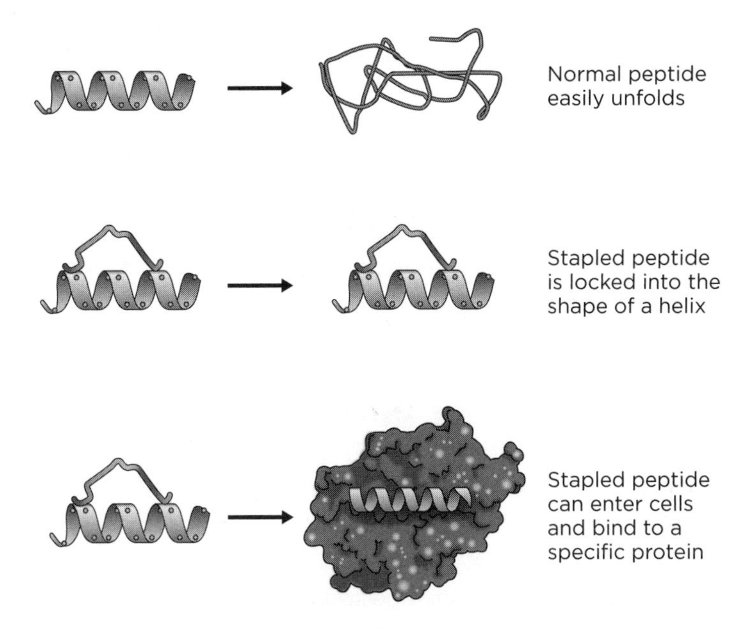

Figure 12.1 Stapled peptides. Peptides can be attached to a chemical staple that holds them in the right shape, making them into potential drug molecules.

In the mid-1990s Verdine began to wonder if there was a way that a chemist could improve upon peptides, making them into drugs. He knew all of the issues associated with peptides, including their inability to penetrate into cells, their susceptibility to degradation by protease enzymes, and their inability to maintain a stable, folded structure. Researchers had been making progress on the problems of cell permeability and stability. Some researchers had also begun exploring ways to stabilize peptides in active conformations.

Working with Christian Schafmeister, a postdoctoral scientist in his laboratory, Verdine built upon these earlier studies attempting to turn peptides into therapeutics. Verdine and Schafmeister designed a means of locking peptides into a single shape—they referred to this as a hydrocarbon staple (see Figure 12.1). They attached a greasy tether linking one part of the peptide with another. When this tether, or staple, was introduced in a suitable way, the peptide was constrained to adopt the shape of an alpha helix. As the peptide began to unfold, the staple prevented any movement, locking it into place.

Shortly after Verdine and Schafmeister reported these results in the *Journal of the American Chemical Society*, they teamed up with Stanley Korsmeyer at the Dana-Farber Cancer Institute to apply this new peptide stapling approach to the design of an inhibitor of the intractable protein BCL-2.

TACKLING THE ANTI-DEATH PROTEIN

Stanley Korsmeyer grew up on his family's farm in Illinois. Unlike others in his family, he decided to pursue a career in biomedical research. As a medical student, he published a paper on which he was the first author (a coveted position for rising scientists) in the *New England Journal of Medicine*. He then did his residency training at the University of California at San Francisco, where he met his future wife.

Working in Tom Waldmann's laboratory at the National Institutes of Health, he collaborated with the geneticist Phil Leder in studying genetic defects present in lymphoma patients. In a study reminiscent of the work on the Philadelphia chromosome that led to the discovery of the BCR-ABL oncogene, Korsmeyer, along with other groups, found a specific chromosomal translocation present in follicular lymphoma cells. This chromosomal aberration involved the unnatural joining of chromosomes 14 and 18, leading to a defect in a gene named B-cell lymphoma-2 (BCL-2).[20]

Korsmeyer moved to the Washington University School of Medicine in St. Louis, where he studied the function of this cancer-linked BCL-2 gene. He found that mice harboring the BCL-2 gene developed lymphomas, confirming it was an important oncogene. Surprisingly, Korsmeyer and several other groups found that the effect of the BCL-2 protein was not to cause unregulated cell growth, as they expected, but rather to prevent tumor cells from dying. BCL-2 seemed to function as a powerful anti-death protein.[21] For his pioneering work on this anti-death protein, Korsmeyer was elected to the National Academy of Sciences of the United States at the young age of 45.

In 1998 Korsmeyer moved to the Dana-Farber Cancer Institute to get involved with the translation of this work into cancer therapies. Shortly after this, he began collaborating with Greg Verdine to make a drug that could inhibit the BCL-2 protein.

The BCL-2 protein does not look like traditional enzymes. It is unlike kinases or proteases, with their large active sites that can bind to small molecules. BCL-2 was considered a challenging, undruggable, target. Further mechanistic studies on BCL-2 revealed that it acts in part by binding to a pro-death protein called BAX. Thus BCL-2 is an example of the many proteins that exert their cellular functions through protein-protein interactions—the challenging type of target for small molecule drugs.

Detailed study of the interaction of BCL-2 with BAX and related proteins showed that there was a fairly well defined peptide at the heart of this interaction. Just as in the case of p53-MDM2, hope emerged that perhaps this type of protein-protein interaction was the sort that could be amenable to disruption with drugs.

Several small molecules have been found that can disrupt this interaction, including one by Stephen Fesik at Abbott Laboratories, using his fragment-based screening approach.[22] However, these small molecules have yet to make it through the drug approval process, so it has not yet been determined whether BCL-2 is truly a druggable target in terms of its susceptibility to a small molecule drug that is suitable for clinical use. Before these small molecule inhibitors had been published, Korsmeyer and Verdine asked whether they could make a different sort of drug, based on the concept of stapled peptides. They created a peptide with a hydrocarbon staple that was predicted to disrupt the BCL-2 interaction with BAX, and thereby allow cell death to proceed normally in tumor cells. The effect would be as if tumor cells did not express high concentrations of the BCL-2 anti-death protein, causing these tumor cells to die.

Korsmeyer and Verdine recruited a young cancer researcher, Loren Walensky, to join this high-risk project. Together, they created a series of stapled peptide molecules and tested them for their ability to shut down the BCL-2 oncoprotein. They hoped that such a stapled peptide would stably assume the correct conformation needed for blocking BCL-2 function. They also hoped that the modified peptide would be stable, and not rapidly destroyed by protease enzymes in the way that unstapled peptides are destroyed. But they had little hope that this peptide would be able to get across the impenetrable barrier of the cell membrane.

Before long, they found a peptide that could block BCL-2 function in a test tube experiment. The surprise was that when they began to study the properties of this new peptide, they were intrigued to see that the hydrocarbon staple on the peptide had solved all of the problems usually associated with peptides. The stapled peptide correctly assumed the right shape, was resistant to destruction, and quite unexpectedly, was able to penetrate into cells.

In a series of experiments, Walensky, Verdine, and Korsmeyer showed that this peptide was also effective in treating mice with tumors. By administering the peptide to mice, they could reduce tumor burden in the treated mice, suggesting this stapled peptide might be a reasonable candidate for developing into a drug. These were exciting results, because they, like the cell-penetrating peptide results, suggested that it might be possible to make large molecule drugs that are effective against proteins that reside inside cells. In other words, it might be possible to step outside of the small molecule box, and tackle undruggable proteins using a wider array of tools than had traditionally been possible.

Given the promise of this peptide stapling technology, in 2005 a new biotechnology company was established to exploit this approach for drug discovery. Within three years this company, Aileron Therapeutics, presented data at a large international cancer meeting showing stapled peptides that were effective in multiple mouse cancer models. The drug development process is a long and arduous one, but there is hope that this new approach will provide drugs against otherwise undruggable proteins. It may be the beginning of a new class of drugs, not only for cancer, but for other diseases as well.

Tragically, Stanley Korsmeyer did not live to see the fruit of this work. He died of the disease he studied for so many years. In March 2005 he died of lung cancer, despite never having smoked. He was 54 years old. His career began by looking for chromosomal aberrations in lymphoma cells and ended by introducing the world to a potentially transformative new type of cancer drug.

THE FUTURE OF ENGINEERED BIOLOGICS

Protein transduction peptides, foldamers, miniproteins, and stapled proteins together represent an emerging class of engineered biologic drugs. These molecules are large—much larger than traditional small molecule drugs. Despite their large size, these molecules can be engineered to be cell-permeable, stably folded, and resistant to degrading enzymes. In other words, these new advances in designing peptides, proteins, and related molecules with novel features potentially enable the creation of biologic drugs that can act inside of cells. This raises the possibility that undruggable proteins might ultimately be conquered by designer biologic drugs. These engineered large molecules may tackle intractable protein-protein interactions and other undruggable proteins. Thus there is a hope, however remote, that eventually all proteins will be targeted by drugs—if not by small molecules, then by engineered biologics.

Moreover, there are other developing technologies that could aid in these efforts. Small RNA molecules are being explored as a means of inactivating messenger RNA molecules. Stem cells are being investigated to replace lost cells in degenerative diseases. Gene therapy is a possible route to introduce new DNA into cells, encoding for the synthesis of new proteins or nucleic acids. These developing technologies offer the promise that intractable diseases and proteins will ultimately succumb to human ingenuity.

13

ACCELERATING THE ARRIVAL OF NEXT-GENERATION DRUGS

There is a grand challenge awaiting us. There are more than 20,000 proteins encoded in the human genome, but only a small fraction—fewer than 15%—are considered druggable given the best current efforts of the scientific community, and only 2% have ever been targeted by a drug. According to the conventional view, this means that there is no hope of making drugs against the vast majority of human proteins. The 85% of proteins that are considered undruggable are understudied and underappreciated as a source of new medicines.

A concerted effort to render these proteins druggable might reap enormous benefits in terms of yielding powerful new medicines for diverse human diseases. By unlocking the potential of these proteins, all manner of pathologies, from cancer to neurodegeneration to infectious diseases, might be brought under control. To provide one specific example, the RAS oncoproteins, despite being studied for over 30 years and being mutated in one out of five human cancers, remain resistant to direct targeting with any drug.

Some researchers are becoming aware of the great benefits that could accrue to society by tackling these undruggable proteins. Indeed, the preceding chapters have described the forefront research that is attempting to solve the problem of making drugs against these undruggable proteins. First, numerous laboratories are trying to create collections of architecturally complex small molecules that might be more effective at binding to these challenging protein families. The creation of these collections frequently makes use of the solid-phase technique invented by Robert Merrifield, and its adaptation to small molecules. The optimal construction of these improved chemical collections will require further discoveries in synthetic chemistry, combinatorial chemistry, and solid-phase synthesis. These next-generation chemical libraries might be one of the keys that unlock the door to the undruggable proteins.

Second, better methods of probing the vast space of possible drug molecules will help solve the problem of undruggable proteins. Fragment-based screening and computational methods for designing drugs that can bind to challenging proteins have great potential to address the issue that finding a good drug molecule from among all the possible molecules is far worse than looking for a needle in a haystack—it is akin to looking for a needle on a planet. Better ways of searching chemistry space may allow for the discovery of effective drugs that simply wouldn't be found by blindly screening preassembled libraries of molecules in the laboratory. However, major advances in computational chemistry and structural biology are needed to make these tools broadly effective.

Third, it might be possible to create a new class of large molecule drugs. The field of protein transduction, with its cell-penetrating peptides, artificial proteins, foldamers, and stapled peptides, seems on the cusp of providing a dramatic change in the type of drugs that are feasible to employ. If this field continues to develop, it might provide the key to accessing many undruggable proteins.

Eventually we will know if these approaches are the right ones for tackling the undruggable proteins. In the course of time, an increasing percentage of proteins will likely be considered druggable. The day may even come, far into the future, when all disease-modifying proteins have been targeted successfully with drugs. This will represent the final triumph of human ingenuity as we reach the pinnacle of medicine—when the majority of possible cures will have been realized.

ACCELERATING THE SEARCH

The crucial question we must ask is: Can anything be done to accelerate the arrival of this future? Surely, the human genome would have been decoded eventually, but the project was dramatically accelerated, perhaps by decades, because of a global, sustained International Human Genome Project. Similarly, a concerted effort to tackle the undruggable proteins could dramatically accelerate progress on this grand challenge for society.

What specifically can be done? At the policy level, political leaders could acknowledge the importance of rendering all proteins druggable and make this a national and international priority. In terms of media, we could see a regularly updated, freely accessible list of the proteins that have been successfully targeted with drugs. I recall seeing a screen in New York City showing the federal debt of the United States in real time, to dramatize the importance of debt control. Why not make available the list of undruggable proteins on a freely accessible Web site, or even display the shrinking number on a sign in Times Square, with excitement mounting as the number continues to fall? In addition, a public service campaign to educate the general population about biology and human health could reap great rewards. I remember watching the *Schoolhouse Rock!* educational musical cartoons that aired on the American Broadcasting System (ABC) in the 1980s; these entertaining bits helped teach kids about science, history, and civics, among other things. A similar effort to educate children about the mechanisms of disease, drugs, and the undruggable proteins would be helpful in spreading awareness of the challenges that prevent medical progress. Moreover, pharmaceutical companies could spend a small part of their advertising budgets to educate the public about basic biomedical science and the challenge of the undruggable proteins. This would ultimately benefit these companies because greater understanding of the monumental challenges of drug discovery would result in greater appreciation of the efforts of the scientists in pharmaceutical companies, and greater celebration of the rare success when a new drug is approved for use in patients. With these broad educational efforts, the size of the challenge of targeting the undruggable proteins might reach the public consciousness.

In terms of coordinated scientific projects, there are many new technologies that could accelerate the discovery of drugs against these challenging proteins. Funding agencies could emphasize the need to discover innovative ways to determine protein structures—despite great progress in this area, it is still challenging to define the atomic-resolution structure of many proteins. Rapid methods for detecting the interactions of molecules are needed, as are better ways of determining which proteins interact with a specific drug. Moreover, the process of optimizing a molecule into a drug is still arduous and haphazard—more systematic approaches would be invaluable. There are many additional technologies and approaches that could be emphasized by funding agencies in a concerted effort to increase the rate of progress against the undruggable proteins.

Finally, at the educational level, the challenge of tackling the undruggable proteins could be introduced into science curricula at the earliest levels. The challenge of the undruggable proteins incorporates many of the key disciplines of the life sciences, rolled up into a fascinating and medically important challenge. By studying this problem students naturally learn chemistry, biochemistry, biophysics, cell biology, and pharmacology. Students would also learn what a protein is, what a drug is, and how proteins function to carry out the processes of life in cells—central aspects of any student's entry into the life sciences.

Science is often taught as a collection of learned facts or formulae, to be memorized and regurgitated. How much more refreshing, for students at every level, it would be to hear about a forefront scientific challenge and some of the new approaches that are being taken. Few people are inherently interested in frogs and beakers, but almost everyone is interested in the health of their families and friends. I imagine many students would be excited by the intellectual challenge of the undruggable proteins and would continue studying science in the hopes of contributing to this grand effort. Hopefully, many of them would be involved in science in the future.

My own educational experience suggests that this problem-centric approach could be effective at motivating students. I recall studying chemistry, biology, and physics in high school; however, the lessons were invariably about balancing or solving equations or memorizing anatomy. A large component of my continued interest in science in the face of this educational paradigm was simply having skilled teachers. However, it wasn't until I went to college that my lifelong interest in science finally clicked. The triggering event was learning organic chemistry and seeing how all of life's processes are simply the interactions and reactions of molecules. I learned that when these molecules are damaged or changed, disease results.

I recall working on a challenging chemistry assignment one evening with great fervor. I was solving a puzzle about how one molecule could be converted into another molecule. With growing enthusiasm I eventually arrived at the answer, working with several friends. A moment of realization dawned on me as I saw how all the questions about disease mechanisms were essentially grander, far more challenging chemistry puzzles waiting to be solved. From detecting the sickle-cell defect in hemoglobin to making drugs that inhibit specific proteins, it became clear that the answers were to be found at the level of molecules. It was in this moment that I knew I wanted to be a part of this majestic search for the answers to molecular puzzles. I wanted to find the molecules that can alter the course of disease and contribute to creating new medicines. My hope is that this excitement over the intellectual challenges and the real-world applications associated with targeting the undruggable proteins will stimulate the next generation of students to enter science. Perhaps by working together we will be able to create that epoch-ending drug against the last of the undruggable proteins.

NOTES

1. THE DRUG DISCOVERY CRISIS

1. Li, J. W. and Vederas, J. C., Drug discovery and natural products: End of an era or an endless frontier? *Science* 325 (5937), 161 (2009); Arlington, S., Accelerating drug discovery: Creating the right environment. *Drug Discov Today* 2 (12), 547 (1997); Drews, J. and Ryser, S., The role of innovation in drug development. *Nat Biotechnol* 15 (13), 1318 (1997); van der Greef, J. and McBurney, R. N., Innovation: Rescuing drug discovery: In vivo systems pathology and systems pharmacology. *Nat Rev Drug Discov* 4 (12), 961 (2005); Czerepak, E. A. and Ryser, S., Drug approvals and failures: Implications for alliances. *Nat Rev Drug Discov* 7, 197 (2008).

2. Swinney, D. C., Biochemical mechanisms of drug action: What does it take for success? *Nat Rev Drug Discov* 3 (9), 801 (2004).

3. Hopkins, A. L. and Groom, C. R., The druggable genome. *Nat Rev Drug Discov* 1 (9), 727 (2002); Russ, A. P. and Lampel, S., The druggable genome: An update. *Drug Discov Today* 10 (23–24), 1607 (2005).

4. Payne, D. J., Gwynn, M. N., Holmes, D. J., and Pompliano, D. L., Drugs for bad bugs: Confronting the challenges of antibacterial discovery. *Nat Rev Drug Discov* 6 (1), 29 (2007).

5. Stockwell, B. R., Haggarty, S. J., and Schreiber, S. L., High-throughput screening of small molecules in miniaturized mammalian cell-based assays involving post-translational modifications. *Chem Biol* 6 (2), 71 (1999); Stockwell, B. R., Hardwick, J. S., Tong, J. K., and Schreiber, S. L., Chemical genetic and

genomic approaches reveal a role for copper in specific gene activation. *J Am Chem Soc* 121, 10662 (1999).

6. Sashidhara, K. V., White, K. N., and Crews, P., A selective account of effective paradigms and significant outcomes in the discovery of inspirational marine natural products. *J Nat Prod* 72 (3), 588 (2009).

7. Verdine, G. L. and Walensky, L. D., The challenge of drugging undruggable targets in cancer: Lessons learned from targeting BCL-2 family members. *Clin Cancer Res* 13 (24), 7264 (2007).

8. Kain, K., The future of cancer therapy: An interview with Gerard Evan. *Dis Model Mech* 1 (2–3), 90 (2008).

9. Buchdunger, E., Zimmermann, J., Mett, H., Meyer, T., Muller, M., Druker, B. J., and Lydon, N. B., Inhibition of the Abl protein-tyrosine kinase in vitro and in vivo by a 2-phenylaminopyrimidine derivative. *Cancer Res* 56 (1), 100 (1996).

10. Capdeville, R., Buchdunger, E., Zimmermann, J., and Matter, A., Glivec (STI571, imatinib), a rationally developed, targeted anticancer drug. *Nat Rev Drug Discov* 1 (7), 493 (2002); Druker, B. J., STI571 (Gleevec) as a paradigm for cancer therapy. *Trends Mol Med* 8 (4), S14 (2002).

11. Osarogiagbon, U. R. and McGlave, P. B., Chronic myelogenous leukemia. *Curr Opin Hematol* 6 (4), 241 (1999).

12. Nowell, P. C., Discovery of the Philadelphia chromosome: A personal perspective. *J Clin Invest* 117 (8), 2033 (2007).

13. Sherbenou, D. W. and Druker, B. J., Applying the discovery of the Philadelphia chromosome. *J Clin Invest* 117 (8), 2067 (2007).

14. Weinstein, I. B. and Joe, A., Oncogene addiction. *Cancer Res* 68 (9), 3077 (2008); Weinstein, I. B. and Joe, A. K., Mechanisms of disease: Oncogene addiction—a rationale for molecular targeting in cancer therapy. *Nat Clin Pract Oncol* 3 (8), 448 (2006); Weinstein, I. B., Cancer. Addiction to oncogenes—the Achilles heal of cancer. *Science* 297 (5578), 63 (2002).

15. The Lasker Foundation (2009). http://www.laskerfoundation.org/awards/2009clinical.htm.

16. Buchdunger et al., Inhibition of the Abl protein-tyrosine kinase; Druker, B. J., Perspectives on the development of imatinib and the future of cancer research. *Nat Med* 15 (10), 1149 (2009).

17. Buchdunger, E., Zimmermann, J., Mett, H., Meyer, T., Muller, M., Regenass, U., and Lydon, N. B., Selective inhibition of the platelet-derived growth factor signal transduction pathway by a protein-tyrosine kinase inhibitor of the 2-phenylaminopyrimidine class. *Proc Natl Acad Sci USA* 92 (7), 2558 (1995).

18. Druker, B. J., Tamura, S., Buchdunger, E., Ohno, S., Segal, G. M., Fanning, S., Zimmermann, J., and Lydon, N. B., Effects of a selective inhibitor of the Abl tyrosine kinase on the growth of Bcr-Abl positive cells. *Nat Med* 2 (5), 561 (1996).

19. Druker, STI571 (Gleevec) as a paradigm for cancer therapy.

20. Dietel, M., *Targeted Therapies in Cancer*. Berlin and New York: Springer, 2007.

2. A NEW SCIENCE OF MOLECULES

1. Brock, W. H., *Justus von Liebig: The Chemical Gatekeeper*. Cambridge and New York: Cambridge University Press, 2002, p. 2.

2. Brown, J. C., *A History of Chemistry from the Earliest Times*. Philadelphia: P. Blakiston's Son & Co., 1920, p. 369.

3. Brock, *Justus von Liebig*, p. 6.

4. Ibid.

5. Ibid., p. 7.

6. Ibid., p. 34.

7. Brown, *A History of Chemistry from the Earliest Times*, p. 369.

8. Brock, *Justus von Liebig*, p. 83.

9. Pershad, J., Palmisano, P., and Nichols, M., Chloral hydrate: The good and the bad. *Pediatr Emerg Care* 15 (6), 432 (1999).

10. Weight, F. F., Aguayo, L. G., White, G., Lovinger, D. M., and Peoples, R. W., GABA- and glutamate-gated ion channels as molecular sites of alcohol and anesthetic action. *Adv Biochem Psychopharmacol* 47, 335 (1992); Istaphanous, G. K. and Loepke, A. W., General anesthetics and the developing brain. *Curr Opin Anaesthesiol* 22 (3), 368 (2009).

11. Brock, *Justus von Liebig*, p. 75.

12. Kurzer, F., Fulminic acid in the history of organic chemistry. *J Chem Educ* 77 (7), 851 (2000).

13. Brock, *Justus von Liebig*, p. 72.

14. Kurzer, Fulminic acid in the history of organic chemistry.

15. Ibid.

16. Valentin, J., *Friedrich Wohler*. Stuttgart: Wissenschaftliche Verlagsgesellschaft M.B.H., 1949.

17. Kurzer, Fulminic acid in the history of organic chemistry.

18. Lundgren, A., *Berzelius och den kemiska atomteorin*. Uppsula: Almqvist & Wiksell International distribution, 1979.

19. Gibson, D. G., Glass, J. I., Lartigue, C., Noskov, V. N., Chuang, R. Y., Algire, M. A., Benders, G. A., Montague, M. G., Ma, L., Moodie, M. M., Merryman, C., Vashee, S., et al., Creation of a bacterial cell controlled by a chemically synthesized genome. *Science* 329 (5987), 52 (2010).

20. Wermuth, C. G., *The Practice of Medicinal Chemistry*, 3rd ed. New York: Academic Press, 2008, p. 6.

21. Ibid.

22. Stockwell, B. R., Chemical genetics: Ligand-based discovery of gene function. *Nat Rev Genet* 1 (2), 116 (2000); Schmitz, R., Friedrich Wilhelm Serturner and the discovery of morphine. *Pharm Hist* 27 (2), 61 (1985).

23. Schmitz, Friedrich Wilhelm Serturner and the discovery of morphine.

24. Ibid.

25. Lipinski, C. A., Lombardo, F., Dominy, B. W., and Feeney, P. J., Experimental and computational approaches to estimate solubility and permeability in drug discovery and development settings. *Adv Drug Deliv Rev* 46 (1–3), 3 (2001); Lipinski, C. A., Drug-like properties and the causes of poor solubility and poor permeability. *J Pharmacol Toxicol Methods* 44 (1), 235 (2000).

26. Lipinski et al., Experimental and computational approaches to estimate solubility and permeability.

27. Chorghade, M. S., *Drug Discovery and Development. Volume 1: Drug Discovery*. Hoboken, NJ: John Wiley & Sons, 2006.

28. Booth, M., *Opium: A History*. New York: St. Martin's Press, 1999, p. 73.

29. Smith, S. L., *Heroin*. New York: The Rosen Publishing Group, 1995, p. 13.

30. Mahdi, J. G., Mahdi, A. J., and Bowen, I. D., The historical analysis of aspirin discovery, its relation to the willow tree and antiproliferative and anticancer potential. *Cell Prolif* 39 (2), 147 (2006).

31. Ibid.

32. Ibid.

33. Miner, J. and Hoffhines, A., The discovery of aspirin's antithrombotic effects. *Tex Heart Inst J* 34 (2), 179 (2007).

34. Hawkey, C. J., COX-2 chronology. *Gut* 54 (11), 1509 (2005).

35. Mahdi, The historical analysis of aspirin discovery.

36. Ibid.

37. Snyder, S. H., Pert, C. B., and Pasternak, G. W., The opiate receptor. *Ann Intern Med* 81 (4), 534 (1974); Kuhar, M. J., Pert, C. B., and Snyder, S. H., Regional distribution of opiate receptor binding in monkey and human brain. *Nature* 245 (5426), 447 (1973); Pert, C. B. and Snyder, S. H., Opiate receptor: Demonstration in nervous tissue. *Science* 179 (77), 1011 (1973).

38. Pert, A., Simantov, R., and Snyder, S. H., A morphine-like factor in mammalian brain: Analgesic activity in rats. *Brain Res* 136 (3), 523 (1977); Hughes, J., Smith, T. W., Kosterlitz, H. W., Fothergill, L. A., Morgan, B. A., and Morris, H. R., Identification of two related pentapeptides from the brain with potent opiate agonist activity. *Nature* 258 (5536), 577 (1975).

39. Ingram, V. M., A specific chemical difference between the globins of normal human and sickle-cell anaemia haemoglobin. *Nature* 178 (4537), 792 (1956).

40. Root, D. E., Flaherty, S. P., Kelley, B. P., and Stockwell, B. R., Biological mechanism profiling using an annotated compound library. *Chem Biol* 10 (9), 881 (2003).

41. Blanchard, B. J., Stockwell, B. R., and Ingram, V. M., Eliminating membrane depolarization caused by the Alzheimer peptide Abeta(1-42, aggr.). *Biochem Biophys Res Commun* 293 (4), 1204 (2002).

42. Blanchard, B. J., Chen, A., Rozeboom, L. M., Stafford, K. A., Weigele, P., and Ingram, V. M., Efficient reversal of Alzheimer's disease fibril formation and

elimination of neurotoxicity by a small molecule. *Proc Natl Acad Sci USA* 101 (40), 14326 (2004).

43. Wermuth, *The Practice of Medicinal Chemistry*, p. 18.

44. Ibid., p. 12.

45. Mann, T. and Keilin, D., Sulphanilamide as a specific inhibitor of carbonic anhydrase. *Nature* 146, 164 (1940).

46. Valenstein, E. S., *The War of the Soups and the Sparks: The Discovery of Neurotransmitters and the Dispute Over How Nerves Communicate*. New York: Columbia University Press, 2005, p. 13.

47. Ibid., p. 14.

48. Quirke, V., Putting theory into practice: James Black, receptor theory and the development of the beta-blockers at ICI, 1958–1978. *Med Hist* 50 (1), 69 (2006).

49. Ibid.

50. Greene, W. C., A history of AIDS: Looking back to see ahead. *Eur J Immunol* 37, Suppl. 1, S94 (2007).

51. Ibid.

52. Ibid.

53. Ibid.

54. Ibid.

3. THE BIRTH OF THE FIRST CANCER DRUGS

1. Croddy, E. A., Wirtz, J. J., and Larsen, J. A., *Weapons of Mass Destruction: An Encyclopedia of Worldwide Policy, Technology, and History*. Santa Barbara, CA: ABC-CLIO, 2005, p. 86.

2. Duchovic, R. J., Mustard gas: Its pre–World War I history. *J Chem Ed* 84 (6), 944 (2007).

3. Ibid.

4. Ibid.

5. Ibid.

6. Ibid.

7. Li, J. J., *Triumph of the Heart: The Story of Statins*. Oxford: Oxford University Press, 2009, p. 17.

8. Ibid.

9. Duchovic, Mustard gas.

10. Richter, D., *Chemical Soldiers: British Gas Warfare in World War I*. Lawrence: University Press of Kansas, 1992, p. 198.

11. Coffey, P., *Cathedrals of Science : The Personalities and Rivalries That Made Modern Chemistry*. London and New York: Oxford University Press, 2008, p. 97.

12. Richter, *Chemical Soldiers.*

13. Wattana, M. and Bey, T., Mustard gas or sulfur mustard: An old chemical agent as a new terrorist threat. *Prehosp Disaster Med* 24 (1), 19 (2009).

14. Ibid.

15. Senior, J. K., Manufacture of mustard gas in World War I. *Armed Forces Chemical Journal* 12 (br. 5), 12 (1958); Leake, C. D. and Marsh, D. F., Mechanism of action of ordinary war gases. *Science* 96 (2487), 194 (1942).

16. Wattana and Bey, Mustard gas or sulfur mustard; Smith, K. J. and Skelton, H., Chemical warfare agents: Their past and continuing threat and evolving therapies. *Skinmed* 2 (4), 215 (2003).

17. Gilman, A. and Philips, F. S., The biological actions and therapeutic applications of the B-chloroethyl amines and sulfides. *Science* 103 (2675), 409 (1946).

18. Ibid.

19. Ibid.

20. Ibid.

21. Goodman, L. S., Wintrobe, M. M., Dameshek, W., Goodman, M. J., Gilman, A., and McLennan, M. T., Nitrogen mustard therapy. Use of methyl-bis(beta-chloroethyl)amine hydrochloride and tris(beta-chloroethyl)amine hydrochloride for Hodgkin's disease, lymphosarcoma, leukemia and certain allied and miscellaneous disorders. *JAMA* 132 (3), 126 (1946).

22. Ibid.

23. Ibid.

24. Ibid.

25. Brock, N., The history of the oxazaphosphorine cytostatics. *Cancer* 78 (3), 542 (1996).

26. Ibid.

27. Ibid.

28. Teicher, B. A., *Cancer Therapeutics: Experimental and Clinical Agents.* Totowa, NJ: Humana Press, 1997.

29. Devita, V. T., Jr., Serpick, A. A., and Carbone, P. P., Combination chemotherapy in the treatment of advanced Hodgkin's disease. *Ann Intern Med* 73 (6), 881 (1970).

30. Asou, N., Suzushima, H., Nishimura, S., Okubo, T., Yamasaki, H., Osato, M., Hoshino, K., Takatsuki, K., and Mitsuya, H., Long-term remission in an elderly patient with mantle cell leukemia treated with low-dose cyclophosphamide. *Am J Hematol* 63 (1), 35 (2000).

31. Wolff, J. A., Chronicle: First light on the horizon: The dawn of chemotherapy. *Med Pediatr Oncol* 33 (4), 405 (1999).

32. Heinle, R. W. and Welch, A. D., Experiments with pteroylglutamic acid and pteroylglutamic acid deficiency in human leukemia. *J Clin Invest* 27 (4), 539 (1948).

33. Wolff, Chronicle: First light on the horizon: The dawn of chemotherapy.

34. Ibid.

35. Ibid.

36. Ibid.

37. Ibid.

38. Dy, G. K. and Adjei, A. A., Systemic cancer therapy: Evolution over the last 60 years. *Cancer* 113, Suppl. 7, 1857 (2008).

39. Ringer, S., Concerning the influence exerted by each of the constituents of the blood on the contraction of the ventricle. *J Physiol* 3, 380 (1880–1882).

40. Ibid.

41. Landecker, H., *Culturing Life: How Cells Became Technologies*. London: Harvard University Press, 2007, p. 15.

42. Pantazis, P., Giovanella, B. C., and Rothenberg, M. L., *The Camptothecins: From Discovery to the Patient*. New York: The New York Academy of Sciences, 1996; O'Leary, J. and Muggia, F. M., Camptothecins: A review of their development and schedules of administration. *Eur J Cancer* 34 (10), 1500 (1998).

43. Dy and Adjei, Systemic cancer therapy; Schwartz, G. N., Pendyala, L., Kindler, H., Meropol, N., Perez, R., Raghavan, D., and Creaven, P., The clinical development of paclitaxel and the paclitaxel/carboplatin combination. *Eur J Cancer* 34 (10), 1543 (1998); Hande, K. R., Etoposide: Four decades of development of a topoisomerase II inhibitor. *Eur J Cancer* 34 (10), 1514 (1998).

44. Cragg, G. M., Paclitaxel (Taxol): A success story with valuable lessons for natural product drug discovery and development. *Med Res Rev* 18 (5), 315 (1998).

45. Baguley, B. C. and Kerr, D. J., *Anticancer Drug Development*. San Diego: Academic Press, 2002.

46. Kelland, L., The resurgence of platinum-based cancer chemotherapy. *Nat Rev Cancer* 7 (8), 573 (2007); Lebwohl, D. and Canetta, R., Clinical development of platinum complexes in cancer therapy: An historical perspective and an update. *Eur J Cancer* 34 (10), 1522 (1998).

47. Missailidis, S., *Anticancer Therapeutics*. Chichester, UK: Wiley-Blackwell, 2008; Perry, M. C., *The Chemotherapy Source Book*. Baltimore: Williams & Wilkins, 1992.

48. Jordan, V. C., Tamoxifen (ICI46,474) as a targeted therapy to treat and prevent breast cancer. *Br J Pharmacol* 147, Suppl. 1, S269 (2006).

49. Smith and Skelton, Chemical warfare agents; Jordan, Tamoxifen (ICI46,474) as a targeted therapy; Miller, C. P., SERMs: Evolutionary chemistry, revolutionary biology. *Curr Pharm Des* 8 (23), 2089 (2002); Hughes-Davies, L., Caldas, C., and Wishart, G. C., Tamoxifen: The drug that came in from the cold. *Br J Cancer* 101 (6), 875 (2009).

50. Santen, R. J., Brodie, H., Simpson, E. R., Siiteri, P. K., and Brodie, A., History of aromatase: Saga of an important biological mediator and therapeutic target. *Endocr Rev* 30 (4), 343 (2009).

51. McLeod, D. G., Hormonal therapy: Historical perspective to future directions. *Urology* 61 (2), Suppl. 1, 3 (2003).

52. Missailidis, *Anticancer Therapeutics.*

53. Gimbrone, M. A., Leapman, S. B., Cotran, R. S., and Folkman, J., Tumor dormancy in vivo by prevention of neovascularization. *J Exp Med* 136 (2), 261 (1972).

54. Bielenberg, D. R. and D'Amore, P. A., Judah Folkman's contribution to the inhibition of angiogenesis. *Lymphat Res Biol* 6 (3–4), 203 (2008).

55. Ebos, J. M., Lee, C. R., Cruz-Munoz, W., Bjarnason, G. A., Christensen, J. G., and Kerbel, R. S., Accelerated metastasis after short-term treatment with a potent inhibitor of tumor angiogenesis. *Cancer Cell* 15 (3), 232 (2009); Kerbel, R. S., Tumor angiogenesis. *N Engl J Med* 358 (19), 2039 (2008).

4. A NEW COMPANY CREATING DRUG COMBINATIONS

1. Rotella, D. P., Phosphodiesterase 5 inhibitors: Current status and potential applications. *Nat Rev Drug Discov* 1 (9), 674 (2002).

2. Keith, C. T., Borisy, A. A., and Stockwell, B. R., Multicomponent therapeutics for networked systems. *Nat Rev Drug Discov* 4 (1), 71 (2005).

3. Ibid.

4. Schreiber, S. L., Chemical genetics resulting from a passion for synthetic organic chemistry. *Bioorg Med Chem* 6, 1127 (1998).

5. Root, D. E., Flaherty, S. P., Kelley, B. P., and Stockwell, B. R., Biological mechanism profiling using an annotated compound library. *Chem Biol* 10 (9), 881 (2003).

6. Overington, J. P., Al-Lazikani, B., and Hopkins, A. L., How many drug targets are there? *Nat Rev Drug Discov* 5 (12), 993 (2006).

7. Stockwell, B. R., Hardwick, J. S., Tong, J. K., and Schreiber, S. L., Chemical genetic and genomic approaches reveal a role for copper in specific gene activation. *J Am Chem Soc* 121 (45), 10662 (1999).

8. Borisy, A. A., Elliott, P. J., Hurst, N. W., Lee, M. S., Lehar, J., Price, E. R., Serbedzija, G., Zimmermann, G. R., Foley, M. A., Stockwell, B. R., and Keith, C. T., Systematic discovery of multicomponent therapeutics. *Proc Natl Acad Sci USA* 100 (13), 7977 (2003).

9. Ibid.; Lehar, J., Zimmermann, G. R., Krueger, A. S., Molnar, R. A., Ledell, J. T., Heilbut, A. M., Short, G. F., Giusti, L. C., Nolan, G. P., Magid, O. A., Lee, M. S., Borisy, A. A., et al., Chemical combination effects predict connectivity in biological systems. *Mol Syst Biol* 3, 80 (2007); Lehar, J., Stockwell, B. R., Giaever, G., and Nislow, C., Combination chemical genetics. *Nat Chem Biol* 4 (11), 674 (2008).

5. THE UNDRUGGABLE RAS PROTEIN

1. Schwann, T. and Smith, H. (translator), *Microscopical Researches into the Accordance in the Structure and Growth of Animals and Plants*. London: The Sydenham Society, 1847.

2. Vasil, I. K., A history of plant biotechnology: From the cell theory of Schleiden and Schwann to biotech crops. *Plant Cell Rep* 27 (9), 1423 (2008); Weinberg, R. A., *The Biology of Cancer*. New York: Garland Science, 2007, p. 27.

3. Griffith, F., The significance of Pneumococcal types. *J Hyg (Lond)* 27 (2), 113 (1928).

4. Vasil, A history of plant biotechnology; Avery, O. T., Macleod, C. M., and McCarty, M., Studies on the chemical nature of the substance inducing transformation of pneumococcal types. *J Exp Med* 79 (2), 137 (1944).

5. Javier, R. T. and Butel, J. S., The history of tumor virology. *Cancer Res* 68 (19), 7693 (2008).

6. Temin, H. M. and Rubin, H., Characteristics of an assay for Rous sarcoma virus and Rous sarcoma cells in tissue culture. *Virology* 6 (3), 669 (1958); Rubin, H. and Temin, H. M., Infection with the Rous sarcoma virus in vitro. *Fed Proc* 17 (4), 994 (1958).

7. Weinberg, *The Biology of Cancer*, p. 63.

8. Toyoshima, K. and Vogt, P. K., Temperature sensitive mutants of an avian sarcoma virus. *Virology* 39 (4), 930 (1969); Martin, G. S., Rous sarcoma virus: A function required for the maintenance of the transformed state. *Nature* 227 (5262), 1021 (1970); Duesberg, P. H. and Vogt, P. K., Differences between the ribonucleic acids of transforming and nontransforming avian tumor viruses. *Proc Natl Acad Sci USA* 67 (4), 1673 (1970); Coffin, J. M., Hughes, S. H., and Varmus, H. E., *Retroviruses*. Plainview, NY: Cold Spring Harbor Laboratory Press, 1997.

9. Stehelin, D., Varmus, H. E., Bishop, J. M., and Vogt, P. K., DNA related to the transforming gene(s) of avian sarcoma viruses is present in normal avian DNA. *Nature* 260 (5547), 170 (1976).

10. Weinberg, *The Biology of Cancer*, p. 77.

11. Karnoub, A. E. and Weinberg, R. A., Ras oncogenes: Split personalities. *Nat Rev Mol Cell Biol* 9 (7), 517 (2008); DeFeo, D., Gonda, M. A., Young, H. A., Chang, E. H., Lowy, D. R., Scolnick, E. M., and Ellis, R. W., Analysis of two divergent rat genomic clones homologous to the transforming gene of Harvey murine sarcoma virus. *Proc Natl Acad Sci USA* 78 (6), 3328 (1981); Ellis, R. W., DeFeo, D., Furth, M. E., and Scolnick, E. M., Mouse cells contain two distinct ras gene mRNA species that can be translated into a p21 onc protein. *Mol Cell Biol* 2 (11), 1339 (1982).

12. Parada, L. F., Tabin, C. J., Shih, C., and Weinberg, R. A., Human EJ bladder carcinoma oncogene is homologue of Harvey sarcoma virus ras gene. *Nature* 297 (5866), 474 (1982); Der, C. J., Krontiris, T. G., and Cooper, G. M., Transforming genes of human bladder and lung carcinoma cell lines are homologous to the ras genes of Harvey and Kirsten sarcoma viruses. *Proc Natl Acad Sci USA* 79 (11), 3637 (1982); Santos, E., Tronick, S. R., Aaronson, S. A., Pulciani, S., and Barbacid, M., T24 human bladder carcinoma oncogene is an activated form of the normal human homologue of BALB- and Harvey-MSV transforming genes. *Nature* 298 (5872), 343 (1982).

13. Shih, T. Y., Papageorge, A. G., Stokes, P. E., Weeks, M. O., and Scolnick, E. M., Guanine nucleotide-binding and autophosphorylating activities associated with the p21src protein of Harvey murine sarcoma virus. *Nature* 287 (5784), 686 (1980).

14. Rodriguez-Viciana, P., Tetsu, O., Oda, K., Okada, J., Rauen, K., and McCormick, F., Cancer targets in the Ras pathway. *Cold Spring Harb Symp Quant Biol* 70, 461 (2005).

15. Ibid.

16. Basso, A. D., Mirza, A., Liu, G., Long, B. J., Bishop, W. R., and Kirschmeier, P., The farnesyl transferase inhibitor (FTI) SCH66336 (lonafarnib) inhibits Rheb farnesylation and mTOR signaling. Role in FTI enhancement of taxane and tamoxifen anti-tumor activity. *J Biol Chem* 280 (35), 31101 (2005).

17. Rodriguez-Viciana et al., Cancer targets in the Ras pathway.

6. THE DRUGGABLE GENOME

1. Edelson, E., *Gregor Mendel and the Roots of Genetics*. New York: Oxford University Press, 1999, p. 26.

2. Ibid., p. 35.

3. Orel, V., *Gregor Mendel: The First Geneticist*. New York: Oxford University Press, 1996, p. 33.

4. Edelson, *Gregor Mendel and the Roots of Genetics*, p. 45.

5. Ibid., p. 46.

6. Fairbanks, D. J., A century of genetics. In *Shrubland Ecosystem Genetics and Biodiversity: Proceedings*, McArthur, E. D. and Fairbanks, D. J. (editors). Provo, UT: U.S. Department of Agriculture, Forest Service, Rocky Mountain Research Station, Vol. RMRS-P-21, 2001, p. 42.

7. Ibid.

8. Edelson, *Gregor Mendel and the Roots of Genetics*, p. 74.

9. Ibid., p. 80.

10. Allen, G. E., *Thomas Hunt Morgan*. Princeton: Princeton University Press, 1978, p. 4.

11. Ibid., p. 97.

12. Ibid., p. 129.

13. Wenrich, D. H., Clarence Erwin McClung 1870–1946. *Science* 103 (2679), 551 (1946).

14. McClung, C. E., Notes on the accessory chromosome. *Anat Anz* 20, 220 (1901).

15. Crow, E. W. and Crow, J. F., 100 years ago: Walter Sutton and the chromosome theory of heredity. *Genetics* 160 (1), 1 (2002).

16. Stevens, N. M., Studies in spermatogenesis with especial reference to the "accessory chromosome." *Carnegie Inst Washington Publ* 36, 1 (1905).

17. Allen, *Thomas Hunt Morgan*, p. 130.

18. Ibid., p. 142.

19. Shine, I. and Wrobel, S., *Thomas Hunt Morgan: Pioneer of Genetics*. Lexington: University Press of Kentucky, 1976, p. 64.

20. Green, M. M., 2010: A century of drosophila genetics through the prism of the white gene. *Genetics* 184 (1), 3 (2010).

21. Shine and Wrobel, *Thomas Hunt Morgan*, p. 70.

22. Sturtevant, A. H., *A History of Genetics*. Cold Spring Harbor, NY: Cold Spring Harbor Laboratory Press, 2001.

23. Dunn, L. C., *Genetics in the 20th Century: Essays on the Progress of Genetics During Its First 50 Years*. New York: Macmillan, 1951.

24. Comfort, N. C., *The Tangled Field: Barbara McClintock's Search for the Patterns of Genetic Control*. Cambridge, MA,: Harvard University Press, 2001.

25. Ibid.

26. Wallace, B., *The Search for the Gene*. Ithaca, NY: Cornell University Press, 1992.

27. Ibid.

28. Gaudilliere, J.-P. and Rheinberger, H.-J., *From Molecular Genetics to Genomics: The Mapping Cultures of Twentieth-Century Genetics*. London: Routledge, 2004.

29. Lander, E. S. and Weinberg, R. A., Genomics: Journey to the center of biology. *Science* 287 (5459), 1777 (2000).

30. Fairbanks, D. J., A century of genetics.

31. Ibid.

32. Lander and Weinberg, Genomics.

33. Ibid.

34. The Huntington's Disease Collaborative Research Group, A novel gene containing a trinucleotide repeat that is expanded and unstable on Huntington's disease chromosomes. *Cell* 72 (6), 971 (1993).

35. Gee, H., *Jacob's Ladder: The History of the Human Genome.* New York: W. W. Norton & Company, 2004.

36. Roberts, L., The human genome: Controversial from the start. *Science* 291 (5507), 1182 (2001).

37. Lander and Weinberg, Genomics.

38. Venter, J. C., Adams, M. D., Myers, E. W., Li, P. W., Mural, R. J., Sutton, G. G., Smith, H. O., Yandell, M., Evans, C. A., Holt, R. A., Gocayne, J. D., Amanatides, P., et al., The sequence of the human genome. *Science* 291 (5507), 1304 (2001); Lander, E. S., Linton, L. M., Birren, B., Nusbaum, C., Zody, M. C., Baldwin, J., Devon, K., Dewar, K., Doyle, M., FitzHugh, W., Funke, R., Gage, D., et al., Initial sequencing and analysis of the human genome. *Nature* 409 (6822), 860 (2001); International Human Genome Sequencing Consortium, Finishing the euchromatic sequence of the human genome. *Nature* 431 (7011), 931 (2004).

39. Drews, J., Genomic sciences and the medicine of tomorrow. *Nat Biotechnol* 14 (11), 1516 (1996).

40. Hopkins, A. L. and Groom, C. R., The druggable genome. *Nat Rev Drug Discov* 1 (9), 727 (2002).

41. Overington, J. P., Al-Lazikani, B., and Hopkins, A. L., How many drug targets are there? *Nat Rev Drug Discov* 5 (12), 993 (2006).

42. Russ, A. P. and Lampel, S., The druggable genome: An update. *Drug Discov Today* 10 (23), 1607 (2005).

7. PEERING INSIDE PROTEINS

1. Mattson, J. and Simon, M., *The Story of MRI: The Pioneers of NMR and Magnetic Resonance in Medicine.* Ramat Gan, Israel: Bar-Ilan University Press, 1996, p. 3.

2. Ibid., p. 4.

3. Bragg, W. H. and Bragg, W. L., *X Rays and Crystal Structure*, 5th ed. London: G. Bell, 1925, p. 1.

4. Hausmann, R., *To Grasp the Essence of Life: A History of Molecular Biology.* Dordrecht and Boston: Kluwer Academic Publishers, 2002, p. 18.

5. Bragg and Bragg, *X Rays and Crystal Structure*, p. 3.

6. Ferry, G., *Max Perutz and the Secret of Life.* Cold Spring Harbor, NY: Cold Spring Harbor Laboratory Press, 2008, p. 30.

7. Tanford, C. and Reynolds, J. A., *Nature's Robots: A History of Proteins.* Oxford and New York: Oxford University Press, 2001, p. 112.

8. Perutz, M. F., *Protein Structure: New Approaches to Disease and Therapy.* New York: W. H. Freeman and Co., 1992, p. 1.

9. McPherson, A., A brief history of protein crystal growth. *J. Cryst Growth* 110, 1 (1991); Tanford and Reynolds, *Nature's Robots*, p. 24.

10. Borek, E., *The Atoms Within Us*. New York: Columbia University Press, 1962, p. 116; Tanford and Reynolds, *Nature's Robots*, p. 40.

11. Stretton, A. O., The first sequence: Fred Sanger and insulin. *Genetics* 162 (2), 527 (2002); Sanger, F., The early days of DNA sequences. *Nat Med* 7 (3), 267 (2001).

12. Ferry, *Max Perutz and the Secret of Life*, p. 32.

13. Ferry, G., *Dorothy Hodgkin: A Life*. Cold Spring Harbor, NY: Cold Spring Harbor Laboratory Press, 2000.

14. Perutz, M. F., *Science Is Not a Quiet Life: Unravelling the Atomic Mechanism of Haemoglobin*. London: Imperial College Press; River Edge, NJ: World Scientific, 1997, p. xvii.

15. Ferry, *Max Perutz and the Secret of Life*, p. 125.

16. Semenza, G. and Jaenicke, R., *Selected Topics in the History of Biochemistry: Personal Recollections. VI*. Amsterdam and New York: Elsevier, 2000, p. 5.

17. Ferry, *Max Perutz and the Secret of Life*, p. 129.

18. Hunter, Graeme K., *Vital Forces: The Discovery of the Molecular Basis of Life*. San Diego: Academic Press, 2000, p. 211.

19. Hager, T., *Linus Pauling and the Chemistry of Life*. Oxford and New York: Oxford University Press, 1998, p. 92.

20. Kay, L. E., *The Molecular Vision of Life: Caltech, the Rockefeller Foundation, and the Rise of the New Biology*. Oxford and New York: Oxford University Press, 1993, p. 263; Serafini, A., *Linus Pauling: A Man and His Science*. New York: Paragon House, 1989, p. 131.

21. Hager, *Linus Pauling and the Chemistry of Life*, p. 47.

22. Ferry, *Max Perutz and the Secret of Life*, p. 145.

23. Ibid., p. 147.

24. Wilkins, M. H., Seeds, W. E., Stokes, A. R., and Wilson, H. R., Helical structure of crystalline deoxypentose nucleic acid. *Nature* 172 (4382), 759 (1953); Franklin, R. E. and Gosling, R. G., Evidence for 2-chain helix in crystalline structure of sodium deoxyribonucleate. *Nature* 172 (4369), 156 (1953); Watson, J. D. and Crick, F. H., Molecular structure of nucleic acids: A structure for deoxyribose nucleic acid. *Nature* 171 (4356), 737 (1953).

25. Ferry, *Max Perutz and the Secret of Life*, p. 152; Lee, R., *The Eureka! Moment: 100 Key Scientific Discoveries of the 20th Century*. New York: Routledge, 2002, p. 116.

26. Ferry, *Max Perutz and the Secret of Life*, p. 159.

27. Kendrew, J. C., Bodo, G., Dintzis, H. M., Parrish, R. G., Wyckoff, H., and Phillips, D. C., A three-dimensional model of the myoglobin molecule obtained by x-ray analysis. *Nature* 181 (4610), 662 (1958); Perutz, M. F., Rossmann, M. G., Cullis, A. F., Muirhead, H., Will, G., and North, A. C., Structure of haemoglobin: A three-dimensional Fourier synthesis at 5.5-A. resolution, obtained by X-ray analysis. *Nature* 185 (4711), 416 (1960).

28. Ferry, *Max Perutz and the Secret of Life*, p. 186.

29. *The Nobel Prize in Physiology or Medicine 1962*. Available at http://nobelprize.org/nobel_prizes/medicine/laureates/1962/ (2010).

30. Mattson and Simon, *The story of MRI*, p. 11.

31. Ibid., p. 24.

32. Ibid., p. 28.

33. Ibid., p. 61.

34. Ibid., p. 69.

35. Chodorow, M., *Felix Bloch and Twentieth-Century Physics: Dedicated to Felix Bloch on the Occasion of His Seventy-Fifth Birthday*. Houston: William Marsh Rice University Press, 1980, p. vi.

36. Ibid., p. vii.

37. Ibid., p. viii.

38. Saunders, M., Wishnia, A., and Kirkwood, J. G., The nuclear magnetic resonance spectrum of ribonuclease. *Journal of the American Chemical Society* 79 (12), 3289 (1957); Locke, D. M., *Enzymes: The Agents of Life*. New York: Crown Publishers, 1969, p. 56.

39. Protein Data Bank, *Welcome to the Worldwide Protein Data Bank*. Available at http://www.wwpdb.org/stats.html (2009).

40. Sharff, A. and Jhoti, H., High-throughput crystallography to enhance drug discovery. *Curr Opin Chem Biol* 7 (3), 340 (2003).

41. Herrick, J. B., Peculiar elongated and sickle-shaped red blood corpuscles in a case of severe anemia. 1910. *Yale J Biol Med* 74 (3), 179 (2001).

42. Marengo-Rowe, A. J., Structure-function relations of human hemoglobins. *Proc Bayl Univ Med Cent* 19 (3), 239 (2006).

43. Pauling, L., Itano, H. A., et al., Sickle cell anemia: A molecular disease. *Science* 110 (2865), 543 (1949).

44. Ingram, V. M., A specific chemical difference between the globins of normal human and sickle-cell anaemia haemoglobin. *Nature* 178 (4537), 792 (1956). Source of quotation: Ingram, V. M., Gene mutations in human haemoglobin: The chemical difference between normal and sickle cell haemoglobin. *Nature* 180 (4581), 326 (1957).

45. Wermuth, *Medicinal Chemistry*.

46. Perutz, *Science Is Not a Quiet Life*, p. 473.

47. Klebe, G., Recent developments in structure-based drug design. *J Mol Med* 78 (5), 269 (2000).

48. Wermuth, *Medicinal Chemistry*, p. 608.

49. Cushman, D. W. and Ondetti, M. A., History of the design of captopril and related inhibitors of angiotensin converting enzyme. *Hypertension* 17 (4), 589 (1991).

50. Ibid.

51. Ondetti, M. A., Rubin, B., and Cushman, D. W., Design of specific inhibitors of angiotensin-converting enzyme: New class of orally active antihypertensive agents. *Science* 196 (4288), 441 (1977).

52. Acharya, K. R., Sturrock, E. D., Riordan, J. F., and Ehlers, M. R., Ace revisited: A new target for structure-based drug design. *Nat Rev Drug Discov* 2 (11), 891 (2003).

53. Baldwin, J. J., Ponticello, G. S., Anderson, P. S., Christy, M. E., Murcko, M. A., Randall, W. C., Schwam, H., Sugrue, M. F., Springer, J. P., Gautheron, P., Grove, J., Mallorga, P., et al., Thienothiopyran-2-sulfonamides: Novel topically active carbonic anhydrase inhibitors for the treatment of glaucoma. *J Med Chem* 32 (12), 2510 (1989).

54. Mincione, F., Scozzafava, A., and Supuran, C. T., The development of topically acting carbonic anhydrase inhibitors as antiglaucoma agents. *Curr Pharm Des* 14 (7), 649 (2008).

55. Wermuth, *Medicinal Chemistry*, p. 609.

56. Böhm, H-J. and Schneider, G., *Protein-Ligand Interactions from Molecular Recognition to Drug Design*. Weinheim: Wiley-VCH, 2003.

57. Koppen, H., Virtual screening—What does it give us? *Curr Opin Drug Discov Devel* 12 (3), 397 (2009).

58. Richards, W. G., Virtual screening using grid computing: The screensaver project. *Nat Rev Drug Discov* 1 (7), 551 (2002).

59. McInnes, C., Virtual screening strategies in drug discovery. *Curr Opin Chem Biol* 11 (5), 494 (2007).

60. Richards, Virtual screening using grid computing.

61. Dukhovich, F. S., *Molecular Recognition: Pharmacological Aspects*. New York: Nova Science Publishers, 2004.

62. Hann, M. M., Leach, A. R., and Harper, G., Molecular complexity and its impact on the probability of finding leads for drug discovery. *J Chem Inf Comput Sci* 41 (3), 856 (2001).

63. Ibid., p. 864.

64. Shuker, S. B., Hajduk, P. J., Meadows, R. P., and Fesik, S. W., Discovering high-affinity ligands for proteins: SAR by NMR. *Science* 274 (5292), 1531 (1996).

65. Blum, L. C. and Reymond, J. L., 970 million druglike small molecules for virtual screening in the chemical universe database GDB-13. *J Am Chem Soc* 131 (25), 8732 (2009).

8. THE NATURE OF INTERACTIONS BETWEEN PROTEINS

1. Frieden, C. and Nichol, L. W., *Protein-Protein Interactions*. New York: Wiley, 1981; Hipfner, D. R. and Cohen, S. M., Connecting proliferation and apoptosis in development and disease. *Nat Rev Mol Cell Biol* 5 (10), 805 (2004).

2. McManus, M. T., Laing, W. A., and Allan, A. C., *Protein-Protein Interactions in Plant Biology*. Sheffield, UK: Sheffield Academic Press, 2002.

3. Pauling, L. and Delbruck, M., The nature of the intermolecular forces operative in biological processes. *Science* 92 (2378), 77 (1940).

4. Kleanthous, C., *Protein-Protein Recognition*. Oxford and New York: Oxford University Press, 2000, p. 1.

5. Ibid., p. 9.

6. Ibid., p. 10.

7. Waldmann, H. and Koppitz, M., *Small Molecule-Protein Interactions*. Berlin and New York: Springer-Verlag, 2003, p. 12.

8. Kleanthous, *Protein-Protein Recognition*, p. 11.

9. Ibid., p. 205.

10. Fu, H., *Protein-Protein Interactions: Methods and Applications*. Totowa, NJ: Humana Press, 2004, p. 4.

11. Zhao, L. and Chmielewski, J., Inhibiting protein-protein interactions using designed molecules. *Curr Opin Struct Biol* 15 (1), 31 (2005).

12. Lane, D. P., Such an obsession. *Cancer Biol Ther* 5 (1), 120 (2006).

13. Varmus, H., *The Art and Politics of Science*. New York: W. W. Norton & Company, 2009., p. 43.

14. Lagnado, J., Society News. New honorary members for the Biochemical Society. *The Biochemist* (The Biochemical Society), December 2004, p. 64.

15. Lane, D. P. and Crawford, L. V., T antigen is bound to a host protein in SV40-transformed cells. *Nature* 278 (5701), 261 (1979).

16. Levine, A. J. and Oren, M., The first 30 years of p53: Growing ever more complex. *Nat Rev Cancer* 9, 749 (2009).

17. Fakharzadeh, S. S., Trusko, S. P., and George, D. L., Tumorigenic potential associated with enhanced expression of a gene that is amplified in a mouse tumor cell line. *EMBO J* 10 (6), 1565 (1991); Momand, J., Zambetti, G. P., Olson, D. C., George, D., and Levine, A. J., The mdm-2 oncogene product forms a complex with the p53 protein and inhibits p53-mediated transactivation. *Cell* 69 (7), 1237 (1992).

18. Kussie, P. H., Gorina, S., Marechal, V., Elenbaas, B., Moreau, J., Levine, A. J., and Pavletich, N. P., Structure of the MDM2 oncoprotein bound to the p53 tumor suppressor transactivation domain. *Science* 274 (5289), 950 (1996).

19. Lane, D. P. and Hall, P. A., MDM2—Arbiter of p53's destruction. *Trends Biochem Sci* 22 (10), 374 (1997).

9. FROM PROTEIN-PROTEIN INTERACTIONS TO PERSONALIZED MEDICINES

1. Noble, R. L., The discovery of the vinca alkaloids—chemotherapeutic agents against cancer. *Biochem Cell Biol* 68 (12), 1344 (1990).

2. Ibid.

3. Gupta, S. and Bhattacharyya, B., Antimicrotubular drugs binding to vinca domain of tubulin. *Mol Cell Biochem* 253 (1–2), 41 (2003).

4. Dumontet, C. and Sikic, B. I., Mechanisms of action of and resistance to antitubulin agents: Microtubule dynamics, drug transport, and cell death. *J Clin Oncol* 17 (3), 1061 (1999).

5. Clackson, T. and Wells, J. A., A hot spot of binding energy in a hormone-receptor interface. *Science* 267 (5196), 383 (1995).

6. Arkin, M. R. and Wells, J. A., Small-molecule inhibitors of protein-protein interactions: Progressing towards the dream. *Nat Rev Drug Discov* 3 (4), 301 (2004).

7. Bogan, A. A. and Thorn, K. S., Anatomy of hot spots in protein interfaces. *J Mol Biol* 280 (1), 1 (1998).

8. Tilley, J. W. , Chen, L. , Fry, D. C., Emerson, S. D., Powers, G. D., Biondi, D., Varnell, T., Trilles, R. , Guthrie, R., Mennona, F., Kaplan, G., LeMahieu, R. A., et al., Identification of a small molecule inhibitor of the IL-2/IL-2Ra receptor interaction which binds to IL-2. *J Am Chem Soc* 119 (32), 7589 (1997).

9. Ibid.

10. Vassilev, L. T., MDM2 inhibitors for cancer therapy. *Trends Mol Med* 13 (1), 23 (2007).

11. van't Veer, L. J. and Bernards, R., Enabling personalized cancer medicine through analysis of gene-expression patterns. *Nature* 452 (7187), 564 (2008).

12. Hershko, A., Lessons from the discovery of the ubiquitin system. *Trends Biochem Sci* 21 (11), 446 (1996).

10. A REVOLUTION IN PEPTIDE SYNTHESIS

1. Felix, A. M., A brief biography of Bruce Merrifield: His life and legacy. *Biopolymers* 90 (3), 158 (2008).

2. Stewart, J. M., Remembering Bruce: The early years. *Biopolymers* 90 (3), 185 (2008).

3. Jukes, T. H., Dilworth Wayne Woolley (1914–1966)—a biographical sketch. *J Nutr* 104 (5), 507 (1974).

4. Merrifield, B., The chemical synthesis of proteins. *Protein Sci* 5 (9), 1947 (1996).

5. Mitchell, A. R., Bruce Merrifield and solid-phase peptide synthesis: A historical assessment. *Biopolymers* 90 (3), 175 (2008).

6. Merrifield, The chemical synthesis of proteins.

7. Mitchell, Bruce Merrifield and solid-phase peptide synthesis.

8. Ibid., p. 179.

9. Ibid.

10. Remembering Bruce Merrifield. Collection of reminiscences arranged by Svetlana Mojsov. *Biopolymers* 90 (3), 245 (2008).

11. Ibid.

12. Grimaldi, D. A., *Amber: Window to the Past.* New York: Harry N. Abrams Publishers in association with the American Museum of Natural History, 1996.

13. Grognot, L., United States Patent No. 906,219 (December 8, 1908).

14. McIntosh, J., "Acrolite": A new synthetic resin. *Industrial and Engineering Chemistry* 19 (1), 111 (1927).

15. Galat, A., Nicotinamide from nicotinonitrile by catalytic hydration. *J Am Chem Soc* 70 (11), 3945 (1948).

16. Ibid., p. 3945.

17. Sucholeiki, I., *High-Throughput Synthesis: Principles and Practices.* New York: Marcel Dekker, 2001.

18. Leznoff, C. C. and Wong, J. Y., The use of polymer supports in organic synthesis: The synthesis of monotrityl ethers of symmetrical diols. *Can J Chem* 50, 2892 (1972).

19. Seeberger, P. H., Automated oligosaccharide synthesis. *Chem Soc Rev* 37 (1), 19 (2008).

20. Mei, H.-Y. and Czarnik, A. W., *Integrated Drug Discovery Technologies.* New York: Marcel Dekker, 2002, p. 396.

21. Frank, R., Heikens, W., Heisterberg-Moutsis, G., and Blocker, H., A new general approach for the simultaneous chemical synthesis of large numbers of oligonucleotides: Segmental solid supports. *Nucleic Acids Res* 11 (13), 4365 (1983).

22. Houghten, R. A., General method for the rapid solid-phase synthesis of large numbers of peptides: Specificity of antigen-antibody interaction at the level of individual amino acids. *Proc Natl Acad Sci USA* 82 (15), 5131 (1985).

23. Geysen, H. M., Meloen, R. H., and Barteling, S. J., Use of peptide synthesis to probe viral antigens for epitopes to a resolution of a single amino acid. *Proc Natl Acad Sci USA* 81 (13), 3998 (1984).

24. Furka, A., Combinatorial chemistry: 20 years on. *Drug Discov Today* 7 (1), 1 (2002).

25. Ibid., p. 2.

26. Ibid., p. 4.
27. Lebl, M., Parallel personal comments on "classical" papers in combinatorial chemistry. *J Comb Chem* 1 (1), 3 (1999).

11. A VAST ARRAY OF DRUG CANDIDATES

1. Ban, T. A., In memory of three pioneers. *Int J Neuropsychopharmacol* 9 (4), 475 (2006).
2. Koerner, B. I., Leo Sternbach. Valium. The father of mother's little helpers. *US News World Rep* 127 (25), 58 (1999).
3. Auwers, K. and von Meyenburg, F., Ueber eine neue Synthese von Derivaten des Isindazoles. *Chem Ber* 24, 2370 (1891).
4. Sternbach, L. H., The benzodiazepine story. *J Med Chem* 22 (1), 1 (1979).
5. Ibid.
6. Evans, B. E., Bock, M. G., Rittle, K. E., DiPardo, R. M., Whitter, W. L., Veber, D. F., Anderson, P. S., and Freidinger, R. M., Design of potent, orally effective, nonpeptidal antagonists of the peptide hormone cholecystokinin. *Proc Natl Acad Sci USA* 83 (13), 4918 (1986).
7. Ibid.
8. Evans, B. E., Rittle, K. E., Bock, M. G., DiPardo, R. M., Freidinger, R. M., Whitter, W. L., Lundell, G. F., Veber, D. F., Anderson, P. S., Chang, R. S., Lotti, V. J., Cerino, D. J., et al., Methods for drug discovery: Development of potent, selective, orally effective cholecystokinin antagonists. *J Med Chem* 31 (12), 2235 (1988).
9. Ibid., p. 2241.
10. Bunin, B. A. and Ellman, J. A., A general and expedient method for the solid-phase synthesis of 1,4-benzodiazepine derivatives. *J Am Chem Soc* 114 (27), 10997 (1992).
11. Johnson, K. M., Chen, X., Boitano, A., Swenson, L., Opipari, A. W., Jr., and Glick, G. D., Identification and validation of the mitochondrial F1Fo-ATPase as the molecular target of the immunomodulatory benzodiazepine Bz-423. *Chem Biol* 12 (4), 485 (2005).
12. Mei, H-Y. and Czarnik, A. W., *Integrated Drug Discovery Technologies.* New York: Marcel Dekker, 2002.
13. Nestler, H. P., Bartlett, P.A., and Still, W.C., A general method for molecular tagging of encoded combinatorial chemistry libraries. *J Org Chem* 59 (17), 4723 (1994).
14. Tan, D. S., Foley, M. A., Shair, M. D., and Schreiber, S. L., Stereoselective synthesis of over two million compounds having structural features both

reminiscent of natural products and compatible with miniaturized cell-based assays. *J Am Chem Soc* 120 (33), 8565 (1998).

15. Burke, M. D., Berger, E. M., and Schreiber, S. L., Generating diverse skeletons of small molecules combinatorially. *Science* 302 (5645), 613 (2003).

16. Welsch, M. E., Snyder, S. A., and Stockwell, B. R., Privileged scaffolds for library design and drug discovery. *Curr Opin Chem Biol* 14 (3), 347 (2010).

12. MOVING OUTSIDE THE SMALL MOLECULE BOX

1. Ingham, P. W., The molecular genetics of embryonic pattern formation in Drosophila. *Nature* 335 (6185), 25 (1988).

2. Gehring, W. J., Homeo boxes in the study of development. *Science* 236 (4806), 1245 (1987).

3. Scott, M. P., Weiner, A. J., Hazelrigg, T. I., Polisky, B. A., Pirrotta, V., Scalenghe, F., and Kaufman, T. C., The molecular organization of the Antennapedia locus of Drosophila. *Cell* 35 (3 Pt 2), 763 (1983).

4. Ibid.; McGinnis, W., Garber, R. L., Wirz, J., Kuroiwa, A., and Gehring, W. J., A homologous protein-coding sequence in Drosophila homeotic genes and its conservation in other metazoans. *Cell* 37 (2), 403 (1984); McGinnis, W., Levine, M. S., Hafen, E., Kuroiwa, A., and Gehring, W. J., A conserved DNA sequence in homoeotic genes of the Drosophila Antennapedia and bithorax complexes. *Nature* 308 (5958), 428 (1984).

5. Joliot, A., Pernelle, C., Deagostini-Bazin, H., and Prochiantz, A., Antennapedia homeobox peptide regulates neural morphogenesis. *Proc Natl Acad Sci USA* 88 (5), 1864 (1991).

6. Derossi, D., Chassaing, G., and Prochiantz, A., Trojan peptides: The penetratin system for intracellular delivery. *Trends Cell Biol* 8 (2), 84 (1998).

7. Lindgren, M., Hallbrink, M., Prochiantz, A., and Langel, U., Cell-penetrating peptides. *Trends Pharmacol Sci* 21 (3), 99 (2000).

8. Frankel, A. D. and Pabo, C. O., Cellular uptake of the tat protein from human immunodeficiency virus. *Cell* 55 (6), 1189 (1988).

9. Mae, M. and Langel, U., Cell-penetrating peptides as vectors for peptide, protein and oligonucleotide delivery. *Curr Opin Pharmacol* 6 (5), 509 (2006).

10. Pooga, M., Soomets, U., Hallbrink, M., Valkna, A., Saar, K., Rezaei, K., Kahl, U., Hao, J. X., Xu, X. J., Wiesenfeld-Hallin, Z., Hokfelt, T., Bartfai, T., et al., Cell penetrating PNA constructs regulate galanin receptor levels and modify pain transmission in vivo. *Nat Biotechnol* 16 (9), 857 (1998); Schwarze, S. R., Ho, A., Vocero-Akbani, A., and Dowdy, S. F., In vivo protein transduction: Delivery of a biologically active protein into the mouse. *Science* 285 (5433), 1569 (1999).

11. Shin, I., Edl, J., Biswas, S., Lin, P. C., Mernaugh, R., and Arteaga, C. L., Proapoptotic activity of cell-permeable anti-Akt single-chain antibodies. *Cancer Res* 65 (7), 2815 (2005).

12. Wender, P. A., Mitchell, D. J., Pattabiraman, K., Pelkey, E. T., Steinman, L., and Rothbard, J. B., The design, synthesis, and evaluation of molecules that enable or enhance cellular uptake: Peptoid molecular transporters. *Proc Natl Acad Sci USA* 97 (24), 13003 (2000).

13. Chen, L., Wright, L. R., Chen, C. H., Oliver, S. F., Wender, P. A., and Mochly-Rosen, D., Molecular transporters for peptides: Delivery of a cardioprotective epsilonPKC agonist peptide into cells and intact ischemic heart using a transport system, R(7). *Chem Biol* 8 (12), 1123 (2001).

14. Bautista, A. D., Craig, C. J., Harker, E. A., and Schepartz, A., Sophistication of foldamer form and function in vitro and in vivo. *Curr Opin Chem Biol* 11 (6), 685 (2007).

15. Cheng, R. P., Beyond de novo protein design—de novo design of nonnatural folded oligomers. *Curr Opin Struct Biol* 14 (4), 512 (2004).

16. Appella, D. H., Christianson, L. A., Karle, I. L., Powell, D. R., and Gellman, S. H., ß-peptide foldamers: Robust helix formation in a new family of ß-amino acid oligomers. *J Am Chem Soc* 118 (51), 13071 (1996).

17. Porter, E. A., Wang, X., Lee, H. S., Weisblum, B., and Gellman, S. H., Nonhaemolytic beta-amino-acid oligomers. *Nature* 404 (6778), 565 (2000).

18. Hill, D. J., Mio, M. J., Prince, R. B., Hughes, T. S., and Moore, J. S., A field guide to foldamers. *Chem Rev* 101 (12), 3893 (2001).

19. Daniels, D. S., Petersson, E. J., Qiu, J. X., and Schepartz, A., High-resolution structure of a beta-peptide bundle. *J Am Chem Soc* 129 (6), 1532 (2007).

20. Ley, T. J., Retrospective: Stanley Joel Korsmeyer (1950–2005). *Science* 308 (5723), 803 (2005).

21. Korsmeyer, S. J., BCL-2 gene family and the regulation of programmed cell death. *Cancer Res* 59 (7 Suppl), 1693s (1999).

22. Lessene, G., Czabotar, P. E., and Colman, P. M., BCL-2 family antagonists for cancer therapy. *Nat Rev Drug Discov* 7 (12), 989 (2008).

GLOSSARY

Acid: A molecule with a removable proton. A proton is a hydrogen atom lacking an electron. When a proton is removed from an acid, it dissolves into the surrounding solution (such as water) and makes the solution acidic.

Amide bond: The type of connection between two amino acids when they are strung together to make peptides or proteins. Also known as a peptide bond.

Amino acids: Simple molecules with a characteristic chemical structure that are stitched together into long sequences to form peptides and proteins.

Angstrom: A unit of length that is about 300 million times smaller than an inch. The size of atoms is conveniently described in angstroms because atoms are in the range of one to three angstroms in diameter.

Catalyze: To cause a reaction to go faster than usual. Enzymes are able to catalyze specific chemical reactions.

Chromosome: An organized package of DNA and protein found inside cells. Genes correspond to particular parts of the DNA molecules that form chromosomes.

DNA: Deoxyribonucleic acid. DNA is formed by stitching together many individual building blocks (nucleotides), which are abbreviated by the letters A, T, G, and C. DNA contains the genetic instructions for passing information from a cell to a daughter cell, or from a parent organism to its offspring.

Drosophila melanogaster: A type of fruit fly used in the science of genetics.

Enzyme: A large molecule, usually a protein, that is able to accelerate a specific chemical reaction.

Farnesyl: A hydrophobic, oily, 15-carbon molecule that is attached to some proteins, such as the RAS proteins. Once attached to a protein, the farnesyl group causes the protein to move to an oily environment, such as the cell membrane.

Farnesylation: The act of adding a farnesyl group onto a protein.

Farnesyltransferase: A protein enzyme that is capable of transferring a farnesyl group onto some proteins, such as RAS.

Hydrophobic: Greasiness, oiliness, or water fearing. Gasoline, lipids, and oils are hydrophobic, which is the opposite of hydrophilic, or water loving.

Hydrophobicity: The degree of greasiness or oiliness.

Imatinib: A drug used to treat patients with chronic myelogenous leukemia (CML). Imatinib inhibits several kinase proteins, including the one present in patients with CML, known as BCR-ABL.

Kinase: An enzyme (almost always a protein) with the ability to transfer a phosphoryl group from ATP onto another molecule, such as another protein. A phosphoryl group is a cluster of four atoms (one phosphorus and three oxygens) that when attached to a protein can change the function of the protein. Thus, kinases change the functions of proteins by attaching phosphoryl groups to them.

Mustard gas: A chemical warfare agent.

Nuclear hormone receptors: Proteins that bind to small molecule hormones, such as vitamins A and D, estrogen, testosterone, and cortisol. Typically, this binding event leads to a series of effects that ultimately cause changes in cellular gene expression, which is how such hormones exert many of their effects on cells.

Oncogene: A gene that can convert a normal cell into a tumor cell.

Phosphoryl group: A cluster of four atoms (one phosphorus and three oxygens) that when attached to a protein can change the function of the protein.

Propanolol: A small molecule drug used to treat hypertension. The drug acts by inhibiting beta-adrenergic receptors and is thus known as a beta-blocker.

Protease: An enzyme (usually a protein) that is able to break down other proteins or peptides by severing the amide (peptide) bonds that connect the individual amino acid units that make up proteins and peptides. Proteases degrade proteins and peptides.

Protein: A type of large molecule present in cells that is built by putting together hundreds of amino acid building blocks. Proteins carry out most of the activities of a typical cell. Most drugs act by binding to specific proteins in or on cells.

Proton: A hydrogen atom lacking an electron. Hydrogen is the simplest of the 92 naturally occurring elements that make up all matter in the universe. These elements occur in their simplest forms as atoms, small particles that can combine to make molecules. An atom of hydrogen is made up of two smaller particles, a proton and an electron. The proton has a positive charge and the electron has a negative charge. Thus, when the electron is removed from a hydrogen atom, all that is left is a positively charged proton.

Proto-oncogene: A normal cellular gene that, when mutated, can be converted into a cancer-causing oncogene.

RAF: A protein that functions as a kinase and is mutated in some cancers.

RALGDS: The RAL guanine nucleotide dissociation stimulator, a protein that regulates the activity of the G protein named RAL, which is similar in structure to RAS. RAL is an abbreviation for RAS-like.

Renal cell carcinoma: A type of cancer of the kidney.

Salt: A substance produced by combining positively and negatively charged molecules. A common example is table salt, formed by combining positively charged sodium (Na^+) and negatively charged chlorine (Cl^-), yielding sodium chloride ($NaCl$).

Sarcoma: A cancer that derives from connective tissue cells, such as muscle, fat, bone, or cartilage.

Small molecule: A molecule that is smaller than a protein, but larger than an atom. Most small molecules consist of a few dozen atoms connected together by specific bonds. Many drugs are small molecules. Small molecules are small enough to fit inside small crevices found on the surface of proteins.

Solubility: The extent to which a material can be completely interspersed (i.e., dissolved) into a liquid. When a molecule is dissolved, it is completely surrounded by molecules of the solvent, such as water.

Sorafenib: A drug approved for the treatment of renal cell carcinoma (kidney cancer) and hepatocellular carcinoma (liver cancer). Sorafenib is marketed under the name Nexavar by the pharmaceutical company Bayer.

Sulfanilamide: A drug used to treat bacterial infections that was later found to increase the rate of urination in patients because of the drug's ability to inhibit the protein carbonic anhydrase.

Trypsinogen: The inactive, precursor form of an enzyme involved in digestion. Trypsinogen can be activated, forming trypsin, which can degrade proteins by breaking them down at specific positions.

Undruggable: Impossible to create a drug against. An undruggable protein is a protein for which it is impossible to make a drug that binds to the protein.

INDEX

Note: Page numbers in *italics* indicate figures.